Contested Environments

The Open University Course Team

Prof. Andrew Blowers, OBE, Faculty of Social Sciences (Course Co-chair and Chair of Block 4)
Dr Steve Hinchliffe, Faculty of Social Sciences (Course Co-chair and Chair of Block 1)
Dr Dick Morris, Faculty of Technology (Chair of Block 2)
Dr Nick Bingham, Faculty of Social Sciences (Chair of Block 3)
Dr Rod Barratt, Faculty of Technology
Dr Chris Belshaw, Faculty of Arts
Roger Blackmore, Faculty of Technology
Dr Mark Brandon, Faculty of Science
Prof. David Elliott, Faculty of Technology
Dr Joanna Freeland, Faculty of Science
Dr Wendy Maples, Faculty of Social Sciences
Dr David Morse, Faculty of Mathematics and Computing
Dr Stephen Peake, Faculty of Technology
Alan Reddish, Consultant Course Team Member
Varrie Scott, Course Manager
Dr Joe Smith, Faculty of Social Sciences
Dr Sandy Smith, Faculty of Science
Dr Charles Turner, Faculty of Science

Other Open University staff

Melanie Bayley, Editor
Pam Berry, Key Compositor
Sophia Braybrooke, Software Designer
Karen Bridge, Software Designer

Sarah Carr, Producer, BBC
Lene Connolly, Print Buying Controller
Michael Deal, QA Tester
Nigel Draper, Media Account Manager
Alison Edwards, Editor
Liz Freeman, Copublishing Adviser
Sarah Gamman, Rights Administrator
Phil Gauron, Producer, BBC
Carl Gibbard, Graphic Designer
Richard Golden, Production & Presentation Administrator
Kate Goodson, Producer, BBC
Celia Hart, Picture Researcher
Susie Hooley, Secretary, Faculty of Social Sciences
Robert Hughes, Software Designer
Lisa MacHale, Producer, BBC
Margaret McManus, Rights Assistant
Michele Marsh, Course Secretary, Faculty of Social Sciences
Lynda Oddy, QA Testing Manager
David Schulman, Producer, BBC
Lynne Slocombe, Editor
Jan Smith, Secretary, Faculty of Social Sciences
Neeru Thakrar, Secretary, Faculty of Social Sciences
Colin Thomas, Lead Software Designer
Howie Twiner, Graphic Artist
Julie Usher, Compositor
Jenny Walker, Producer, BBC
David Wardell, Media Account Manager
Amanda Willett, Series Producer, BBC
Darren Wycherley, Producer, BBC

Consultants

Dr Paul Anand, Lecturer in Economics, The Open University
Prof. Jacquie Burgess, Professor of Geography, University College London
Dr Noel Castree, Reader in Human Geography, University of Manchester
Prof. Michael Drake, Professor Emeritus, Faculty of Social Sciences, The Open University

Pam Furniss, Lecturer in Environmental Systems, Faculty of Technology, The Open University
Prof. Pieter LeRoy, Professor in Political Sciences of the Environment, Nijmegen University
Dr Annie Taylor, Visiting Fellow, Department of Politics, University of Southampton
Dr Karin Verhagen, Junior Lecturer in Political Sciences of the Environment, Nijmegen University

External assessor

Prof. Kerry Turner, CBE, Director of The Centre for Social and Economic Research on the Global Environment, University of East Anglia

Tutor panel

Claire Appleby, Associate Lecturer, The Open University, Cambridge
Dr Ian Coates, Associate Lecturer, The Open University, Bristol
Dr Arwyn Harris, Associate Lecturer, The Open University, Wales

With thanks to
Caitlin Harvey, Course Manager, Faculty of Social Sciences
Dr Dan Weinbren, Course Manager, Faculty of Social Sciences
John Watson, Technician, Earth Sciences

Contested Environments

edited by Nick Bingham, Andrew Blowers
and Chris Belshaw

wiley.com

in association with

The Open
University

This publication forms part of an Open University course U216 *Environment: Change, Contest and Response*. The complete list of texts that make up this course can be found on the back cover. Details of this and other Open University courses can be obtained from the Course Information and Advice Centre, PO Box 724, The Open University, Milton Keynes MK7 6ZS, United Kingdom: tel. +44 (0)1908 653231, e-mail general-enquiries@open.ac.uk

Alternatively, you may visit the Open University website at www.open.ac.uk where you can learn more about the wide range of courses and packs offered at all levels by The Open University.

To purchase a selection of Open University course materials visit the webshop at www.ouw.co.uk, or contact Open University Worldwide, Michael Young Building, Walton Hall, Milton Keynes MK7 6AA, United Kingdom for a brochure. tel. +44 (0)1908 858785; fax +44 (0)1908 858787; e-mail ouwenq@open.ac.uk

First published 2003 by John Wiley & Sons Ltd in association with The Open University

The Open University
Walton Hall
Milton Keynes
MK7 6AA
United Kingdom
www.open.ac.uk

John Wiley & Sons Ltd
The Atrium
Southern Gate
Chichester
PO19 8SQ

www.wileyeurope.com or www.wiley.com

Email (for orders and customer service enquiries): cs-books@wiley.co.uk

Other Wiley editorial offices: John Wiley & Sons Inc., 111 River Street, Hoboken, NJ 07030, USA; Jossey-Bass, 989 Market Street, San Francisco, CA 94103–1741, USA; Wiley-VCH Verlag GmbH, Boschstr. 12, D–69469 Weinheim, Germany; John Wiley & Sons Australia Ltd, 33 Park Road, Milton, Queensland 4064, Australia; John Wiley & Sons (Asia) Pte Ltd, 2 Clementi Loop #02–01, Jin Xing Distripark, Singapore 129809; John Wiley & Sons Canada Ltd, 22 Worcester Road, Etobicoke, Ontario, Canada M9W 1L1.

Library of Congress Cataloging-in-Publication Data

A catalogue record for this book is available from the Library of Congress.

British Library Cataloguing in Publication Data

A catalogue record for this book is available from the British Library.

ISBN 0 470 85000 0

Edited, designed and typeset by The Open University.

Printed by The Bath Press, Glasgow.

1.1

Contents

Series preface

Environment: Change, Contest and Response is a series of four books designed to introduce readers to many of the principal approaches and topics in contemporary environmental debate and study. The books form the central part of an Open University course, which shares its name with the series title. Each book is free-standing and can be used in a wide range of environmental science and geography courses and environmental studies in universities and colleges.

This series sets out ways of exploring environments; it provides the knowledge and skills that enable us to understand the variety and complexity of environmental issues and processes. The ideas and concepts presented in the series help to equip the reader to participate in debates and actions that are crucial to the well-being – indeed the survival – of environments.

The series takes as its common starting point the following. First, environments are socially and physically dynamic and are subject to competing definitions and interpretations. Second, environments change in ways that affect people, places, non-humans and habitats, but in ways that are likely to reflect differing degrees of vulnerability. Third, the unsettled, uncertain and uneven nature of environmental change poses significant challenges for political and scientific institutions.

The series is structured around the core themes of **changing** environments, **contested** environments and environmental **responses**. The first book in the series sets out these themes, and (as their titles indicate) each of the three remaining books takes one of the themes as its main concern.

In developing the series we have observed the rapidly changing world of environmental studies and policies. The series covers a range of topics from biodiversity to climate change, from wind farms to genetically modified organisms, from the critical role of mass media to the measurement of ecological footprints, and from plate tectonics to global markets. Each issue requires insights from a variety of disciplines in the natural sciences, social sciences, arts, technology and mathematics. The books contain chapters by a range of authors from different disciplines who share a commitment to finding common themes and approaches with which to enhance environmental learning. The chapters have been read and commented on by the multidisciplinary team. The result is a series of books that are unique in their degree of interdisciplinarity and complementarity.

It has not been possible to include all the topics that might find a place in a comprehensive coverage of environmental issues. We have chosen instead to use particular examples in order to develop a set of themes, concepts and questions that can be applied in a variety of contexts. It is our intention that as you read these books, individually or as a series, you will find thought-provoking and innovative approaches that will help you to make sense of the issues we cover, and many more besides. From the outset our aim has been to provide you with the equipment necessary for you to become a sceptical observer of, and participant in, environmental issues.

In the series we shall talk of both 'environment' and its plural, 'environments'. But, what do these terms mean? As a working definition we take 'environment' to indicate surroundings, including physical forms and features (land, water, sky) and living species and habitats.

'Environments' signifies spaces (areas, places, ecosystems) that vary in scale and are connected to, are a part of, a global environment. For instance, we may think in terms of different environments, such as the Scottish Highlands, the Sahara Desert or the Siberian taiga (forest). We may think in terms of components of environments, for instance housing, industry or infrastructures of the built environment. The phrase 'natural environments' may be used to emphasize the contribution of non-human organisms and biological, chemical and physical processes to the world's environment.

It is useful to keep in mind four defining characteristics of environment: environment implies surroundings; nature and human society are not separate but interactive; environment relates to both places and processes; and environments are constantly changing. These characteristics are merely a starting point; as we proceed we shall gradually reveal the relationships and interactions that shape environments.

There are several features of the books that are worthy of comment. While each book is self-contained, you will find references to earlier and later material in the series depicted in bold type. Some of the cross-references will also be highlighted in the margins of the text so that you may easily see their relevance to the topic on which you are currently engaged. The margins are also used in places to emphasize terms that are defined for the first time.

Another feature of the books is the interactivity of the writing. You will find questions and activities throughout the chapters. These are included to help you to think about the materials you are studying, to check your learning and understanding, and in some cases to apply what you have learned more widely to issues that arise outside the text. A final feature of the chapters is the summaries that appear at the end of each major section, to help you check that you have understood the main issues that are being discussed.

We wish to thank all the colleagues who have made this series and the Open University course possible. The complete list of names of those responsible for the course appear on an earlier page. Particular thanks go to: our external assessor, Professor Kerry Turner; our editors, Melanie Bayley, Alison Edwards and Lynne Slocombe; our tutor panel, Claire Appleby, Ian Coates and Arwyn Harris; and to the secretaries in the Geography Discipline, Michelle Marsh, Jan Smith, Neeru Thakrar and Susie Hooley, who have all helped in the preparation of the course. Last but not least, our thanks to our course manager, Varrie Scott, whose efficiency and unfailing good humour ensured that the whole project was brought together so successfully.

Andrew Blowers and Steve Hinchliffe
Co-chairs of The Open University Course Team

Introduction

Nick Bingham

In the margin are the questions also listed on the back cover of this book. The fact that you are now reading this Introduction perhaps reflects your interest in them, or at least that they stimulated your curiosity. If so, you share this interest or curiosity with us, the contributors to this book, for it was questions very like these that originally brought us together to write it. Together, because each of the questions and the concerns of which they speak are related by a theme that became the title of this book, the theme of contested environments.

Why are food scares becoming more common? Whose voices count in decisions affecting the landscapes where we live? Will there soon be wars over water? What makes people protest outside international trade meetings?

This immediately raises another question: what exactly do we mean by the phrase 'contested environments'? In one sense we have found this very easy to answer – we simply mean that environments are often the source of disagreements among interested parties. However, when we tried to flesh out the bare bones of this statement, things became far more tricky, and it was tempting to fall into one of the two traps that arise whenever what counts as an environmental contest is considered. The first is to see contestation everywhere with respect to environments, to assume that environmental issues are inherently the source of controversy of some kind or another. Because of the sorts of environmental stories that usually make the headlines, it is not difficult to gain this impression; as in most arenas, a good argument makes good news. However, it is important to recognise that this state of affairs is at least in part a result of the way in which certain agendas play out within the media, and that a great deal of other environmental activities simply go unreported. In particular, it would be a mistake to underestimate the degree to which environmental concerns can bring people together in more or less consensual contexts striving towards a common goal. The second trap is to see contestation only when an environmental issue produces a highly visible dispute with all the associated signs of outright confrontation such as demonstrations or direct action. This again misses something significant about how we deal with our surroundings: the fact that profound differences of opinion can exist between groups and individuals with respect to environmental matters without ever necessarily coming to a head.

In an attempt to avoid such pitfalls we have come to see environmental contests on a continuum. Such a continuum ranges from environmental situations in which the distance between the positions of the various parties involved is mainly hidden or latent (see Chapters Two and Seven), to those in which they are right out in the open and explicit to the extent that that we can properly term them 'conflicts' (see Chapters One and Six). Further situations demonstrate that contest about environmental issues can be remarkably resilient, while yet others show that issues can move through the continuum over time from being relatively uncontested to a subject of considerable conflict or vice versa. Environmental contests, then, can manifest themselves in very different ways. In order to provide ways of navigating through and making sense of this variation, we have organized the book around three sets of structuring elements.

Perhaps the most important is a set of *analytical concepts* (values, power and action) that we use both to reflect and gain purchase on what we see as common features of all environmental contests. These analytical concepts have a dual function. First they act as guides through the chapter material. For example, in Chapter One, a distinction (which runs through the book) is made between power defined as resource and power defined as discourse. The distinction is used in turn to generate questions, that are asked in connection with three different episodes of contest over genetically modified food. The other function of the analytical concepts follows from the first. Having learned in the chapters how you can apply them productively to case studies, the concepts become a set of tools you can take away from the book and use to help you think through other environmental issues that you may come across.

The second of the book's structuring elements is a set of three *key questions*. How and why do contests over environmental issues arise? How are environmental understandings produced? Can environmental inequalities be reduced? Each of the chapter authors responds to these questions in different ways. The general lessons that we can learn from their answers will form the conclusion to the book.

The final structuring element is the *sequence* of the chapters themselves. While taken together they illustrate the variation in environmental contests, their precise order is intended to suggest another way of steering a course though the overall theme of the book. Thus the first two chapters (on the controversy surrounding genetically modified food and the ways we encounter different types of landscape) introduce the key concepts in some detail (with more emphasis on the power–action relationship in Chapter One and more on the influence of values on action in Chapter Two). The next three chapters (on issues of energy provision, water supply and trade relations) then represent examples of contests at different spatial scales from the local (Chapter Three), through the regional (Chapter Four), to the global (Chapter Five). Finally, the last three chapters of the book, while remaining rooted in particular examples, address more thematic issues of environmental contest: the possibilities of environmental justice (Chapter Six); the increasing role of economic valuation in environmental policy making (Chapter Seven); and the part played by the news media in the shaping of environmental issues (Chapter Eight).

What we offer in this book, then, is not any definitive answers to the sorts of questions with which we started. Indeed, following how environmental contests unfold in practice, as we do here, teaches that any simple responses to them are also likely to be simplistic and thus inappropriate. However, as we try to demonstrate, what do exist are ways of taking environmental contests seriously, and it is these that signal to us the best ways forward.

Chapter One

Food fights: on power, contest and GM

Nick Bingham

Contents

1   Introduction: food matters

In the preface to his *Food: A History*, the historian Felipe Fernández-Armesto writes:

> The great press baron, Lord Northcliffe, used to tell his journalists that four subjects could be relied upon for abiding public interest: crime, love, money, and food. Only the last of these is fundamental and universal. Crime is a minority interest, even in the worst-regulated societies. It is possible to imagine an economy without money and reproduction without love but not life without food. Food, moreover, has a good claim to be the world's most important subject. It is what matters to most people most of the time.
>
> (Fernandez-Armesto, 2001, p.xiii)

Certainly, food matters when we are considering environments. After all, as Fernandez-Armesto (2001) puts it, 'we meet the world on our plates' (or whatever else we happen to be eating or drinking from). Tracing how even the humblest ingredient comes to be part of one of our meals can reveal an enormous amount about the ways in which we are connected to different people, places, and things of various kinds. Take, for example, the coffee that sits in the cup on my desk while I am writing this. The following are just a few of the factors involved in ensuring that I get my early morning caffeine boost: the amount and timing of rainfall on small cooperatively-owned coffee tree plantations located on the mountain slopes of northern Peru; the extent of plant loss due to 'pests'; the success of ecologically sensitive farming practices such as cultivating earth-worms and mulch to increase soil fertility; the vagaries of the Cocoa, Sugar, and Coffee Exchange in New York that regulates global coffee markets; the transport, processing, and packaging methods involved in bringing the crop to the UK retail market; the standards required for organic and fairtrade certification; and the amount of money paid to the farmers for meeting those standards (based on information in Whatmore and Thorne, 1997). In other words, the two end points of the process belie the range of activity that takes place in between (Figure 1.1).

Thinking about food in this way – bringing to our attention the multiplicity of factors involved in the journey from production to consumption – also serves to remind us of why an interdisciplinary approach to environmental issues is so important. As with all environmental issues we can approach food in many ways: as a source of nutrition, as the basis of various cuisines, as a commodity, as a tracker of changing tastes and class relations, as a component of cultural identity, as a pawn in many a political strategy, as a major element of ecosystems, and so on. At the end of the day though (and as we will see in this chapter), if we are to understand and respond adequately to the many environmental issues associated with food then we need to adopt an approach that recognizes that food is all of those things and more besides. As with all things environmental, the scientific, the technical and the social are profoundly entangled together.

Figure 1.1 It takes a lot to make a cup of coffee.

Activity 1.1

How much do you know about where your food comes from? Pick just three ingredients from a recent meal and note down everything you can about how they got from field to fork (or fingers). Remember that scientific, technical and social factors are all likely to be involved. Are you surprised by how much or how little you do know?

Don't worry if you weren't able to get very far with this activity. The fact of the matter is that for most people (at least in the North) most of the time, food is something that is more or less taken for granted. Although, as we have seen, foodstuffs can provide a very direct link with the different environments of which we are all part, those connections are not necessarily ones we consider very often unless we are asked to do so. Obtaining, preparing and consuming what we eat are usually habitual activities, part of the routines by which we organize our everyday lives.

For some people some of the time, however, the generalization that food is taken for granted has become less true, especially over the last couple of decades or so. Over that period a number of '**food scares**' have prompted many individuals and groups to ask the sort of question that I have just asked you: where does our food come from? The BSE or 'mad cow disease' crisis in the UK during the early 1990s, the uproar over bovine growth hormone and milk in the USA during the mid 1990s and the resurfacing of concerns over pesticide residues on food in continental Europe during the late 1990s all represent examples of food

food scare

becoming scary with what seemed like increasing regularity. In each case, the connections between food production and food consumption were no longer assumed to be safe and thus unnecessary to consider, and through the efforts of the public, the press, regulatory bodies and others these connections were rendered increasingly visible. The result has been that the food we eat has become increasingly contested.

The rest of this chapter examines in more detail one particular contest about food – the controversy surrounding genetically modified crops. We cannot look at every aspect of this controversy here, or at all the potential consequences for the various humans and non-humans involved. What we can do is explore how applying the book's theme (contest and conflict) and two of its key concepts (power and action) to the debate surrounding genetic modification (GM) helps us to begin to answer some of the book's key questions. As you should recall from the book's Introduction these are:

- How and why do contests over environmental issues arise?
- How are environmental understandings produced?
- Can environmental inequalities be reduced?

By the end of the chapter you should be able to give an answer to the following versions of those key questions, which are the ones we address in this chapter:

- Why has GM food become so controversial?
- How has the operation of power enabled and constrained action in the GM food controversy?
- Can paying attention to the operation of power offer us ways out of the GM controversy?

The first question is addressed mainly in the following section which, through the example of GM food, looks at what is required for environmental contests and conflicts to arise. The second is addressed in detail in Sections 4 to 6, which explore three episodes of conflict over GM food. Finally, the last question is addressed briefly in the Conclusion, in which I give an example of how I have used the concept of power to think through the controversy.

2   GM food becoming controversial

In 1995, a group of social scientists based at Lancaster University published a report in which they forecast that, sooner or later, major problems with the public acceptance of GM food would emerge in the UK (Grove-White et al., 1995). Events seem to have proved them correct, and it is worth asking on what basis they made such an accurate prediction as we begin to answer our first question: why has GM food become so controversial? Effectively what the researchers had done was first to identify the GM food issue as what I shall define as an environmental contest, and second to establish that it bore all the hallmarks of becoming what I shall define as an environmental conflict. Before we get to these definitions though, we should take a look in a bit more detail at what all the fuss was about (Box 1.1).

Box 1.1   Genetic modification

All plants and animals are made up of building blocks called **cells** each of which contains a centre or nucleus (see Figure 1.2). Inside every nucleus there are chain-like arrangements of what is known as **DNA** (see Figure 1.3) organized into structures called chromosomes. **Genes** are short segments of this DNA material, and by studying how these genes act and interact, scientists began during the twentieth century to understand much about the characteristics of living organisms and how these characteristics are passed on from one generation to the next.

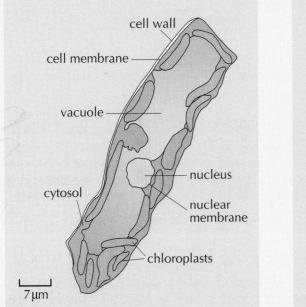

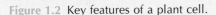

Figure 1.2 Key features of a plant cell.

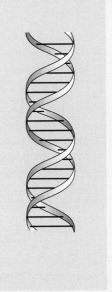

Figure 1.3 The double-helical structure of DNA.

genetic modification

cell

DNA

gene

In the last few decades, a series of technological developments have enabled scientists to move from this activity of studying and understanding towards being able to directly manipulate genes in the laboratory, and it is this process which has became known as genetic engineering or, more commonly at present, genetic modification. The techniques of genetic modification allow scientists to take a gene or genes from one organism and put them into another organism. We have already noted that all plants and animals have genes and certain genes have been associated with certain characteristics or traits displayed by an organism. What genetic modification makes it possible to attempt, then, is to transfer desired traits from one species to another, even species which would not normally breed together.

Genetically modified crops are crops which have been engineered to contain a gene or genes not normally found in that species of plant.

The mechanics of how genetic modification of this sort is actually done with plants involves three main steps:

1   In the first stage, the DNA of the organism which contains the desired trait must be isolated from that organism. Then the gene which is thought to control that trait is identified and cut out of the chain of DNA using a kind of biochemical scissors. At this stage it is cloned, which is to say that multiple identical copies are made.

transgene

2   In the second stage, the gene from the donor organism (or **transgene** as it is called) must be attached to a suitable means of transferring it to the plant which is to receive the desired trait. There are two main ways of doing this. The first is a sort of biological ferry or vector made up of genetic material taken from a chosen bacterium or virus. The second consists of tiny particles or pellets of either gold or tungsten.

3   In the final stage, the transgene is inserted into the recipient plant. If the first method of transfer is used, this is effectively achieved by the vector infecting the recipient plant and in doing so smuggling the transgene into its DNA. If the second method is used, a technology known as a 'gene gun' bombards the recipient plant with the DNA-coated particles in the hope that some may pass through the nucleus and deposit their package of genes in such a way that they are taken up by the recipient plant's DNA. Both these methods at present have extremely low success rates. For this reason, scientists attach what is known as a 'marker gene' to the transgene before attempting to insert it into the recipient plant. After the attempt is made, cells of the recipient plant are grown in a medium containing an antibiotic to which the marker gene is resistant, such that only those which have taken up the new genes are able to survive. These cells are then cultured and grown into mature plants.

Activity 1.2

The description of genetic modification in Box 1.1 is of necessity quite abstract, and thinking it through in relation to a specific example will help you check that you have understood the general process. With this in mind, study Figure 1.4 which shows how genetically modified Bt corn is made and make sure you can identify (a) the donor organism, (b) the recipient organism and (c) what the transgene does. The answers are at the end of the chapter.

environmental contest

What does it mean then that this new technique of food production (amongst other things) should become the subject of contest? In the first instance, not much. What I am calling **environmental contests** can take many forms, ranging from situations where there are few obvious signs of any differences of opinion until those opinions are specifically sought (see Chapter Seven for an example) to those in which contrasting viewpoints on an issue are all too apparent (see Chapter Three for an example). Environmental contests can also take place over many topics – knowledge, diversity, and livelihoods, for example, are all at stake in just the GM food case. The point is that all it takes for an environmental contest to occur is for two or more parties to disagree about something with an environmental dimension. That was certainly the case regarding GM food at the time that the Lancaster researchers mentioned earlier were doing their work. Both the technique itself and its potential implications were made sense of by many of the members of the public who participated in the study in ways that were at odds with those promoting GM. People's attitudes to genetically modified organisms (GMOs) in food products were described in the report as 'ambivalent' overall with some expressing 'significant unease'. Further, confidence in government regulation of food and official assurances of safety was summarized as decidedly 'mixed'.

At that stage, however, such disagreements concerning GM did not amount to a fully fledged example of the particular type of environmental contest that I shall call an **environmental conflict**. Most of the fears about GM foods held by people   environmental conflict in the UK were what the researchers described as 'latent' rather than 'explicit'. In other words, it was not until the public were directly asked what they felt about GM food that such differences in understanding became apparent.

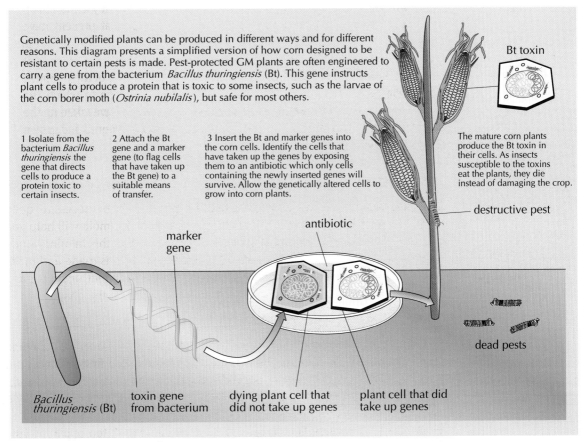

Figure 1.4 The production of Bt corn.
Source: adapted from Zeff, in *Scientific American*, 2001, p.43.

An environmental conflict situation, by contrast, usually has the following four characteristics:

- a very visible public expression of the different understandings and interpretations of the thing or issue in question
- a high degree of commitment to their claims by the different parties involved
- a deployment of power to advance the interests of those parties
- a polarization of the situation into two opposing camps.

The academics who organized the study though were certainly of the opinion that what was latent was soon to become explicit, and they were right in the sense that by 1998 all those ingredients were in place with regard to GM food. This extract from one of the many articles about GM food to appear in the national press both illustrates this nicely and acts as a good summary of the main hopes and fears that surrounded the issue:

> Experts have argued for and against the spread of genetically modified food.
>
> Those for say:
> - genetic engineering can improve the quality of food by making it last longer, improving its flavour
> - it could make food healthier by altering its nutritional content, for example, by reducing levels of fat
> - it could help feed the world by improving yields and reducing costs; fewer pesticides and herbicides will be needed, helping soil conservation; chemicals will be better targeted; less energy will be used in spraying.
>
> Those against say:
> - scientists cannot anticipate potential risks, as BSE shows
> - it will encourage monoculture or crops which are potentially susceptible to disease
> - it concentrates too much power in the hands of agrochemical giants
> - it could lead to a loss of biodiversity
> - consumers are being given no choice, without knowing the consequences
> - the technology poses fundamental ethical questions about life forms
> - there have been disturbing new toxins and allergies related to GM foods.
>
> (*The Guardian*, 4 June 1998, p.15)

This is all very well in terms of getting a sense of what was necessary for GM food to become controversial at a broad level. However, to understand more about how power operated in the resulting controversy, we need to look in more detail at how particular episodes of conflict played out in practice. This will be the focus of Sections 4–6. First, in Section 3, we formulate a set of questions which we will use in our analysis of those episodes.

Summary

- Genetic modification is a technique that allows scientists to take a gene or genes from one organism and put them into another organism.
- Environmental contests can take many forms, be over many things, and only require there to be different understandings or assessments of the issue by the parties involved.
- Environmental conflicts have four characteristics: visible disagreement, a high degree of commitment, the deployment of power, and polarization.

3   Making claims about GM

In their work comparing environmental disputes in France and the USA, a team of sociologists have suggested that there are a number of things that all individuals or groups must do if they want the claims which they make in the public arena to carry any weight. Briefly these comprise: (i) providing a justification for a particular argument, (ii) offering some kind of evidence to back up that argument, and (iii) having institutional, technical, legal or material arrangements in place which support the argumentation under consideration (Thévenot et al., 2000). What I want to do in this section is to expand on what is meant by each of these requirements and to use them as the basis for a number of questions that can be asked of each of the episodes of conflict over GM that we will be analysing later in the chapter. In doing so, I also want to build upon some of the analysis of **power** and **action** introduced by **Hinchliffe and Belshaw (2003)**. The questions we shall ask are as follows.

power

action

What was being contested?

One of the features of the GM issue taken as a whole is how thoroughly it has become controversial. Not only is it difficult to get the many and various parties involved to agree on the facts of the matter, but there does not even exist any consensus about what facts are needed to resolve the issue, or indeed who should be trusted to establish those facts. This does not mean, however, that in every episode of conflict concerning GM food everything is at stake and up for grabs. Rather, as we shall see, individual disputes tend to be characterized by most of the controversy being bracketed off and the disagreement being focused on one particular aspect of the issue.

What justification was given for action?

According to Laurent Thévenot, Michael Moody, and Claudette Lafaye – the sociologists mentioned above whose ideas I am drawing on in this section – all parties involved in public disputes attempt to defend their position by using what they call 'generalized' arguments. By this they mean arguments which, by invoking certain sorts of values, make a claim to a wider applicability than just the immediate situation. They give the examples of 'equality', 'tradition', 'the free market' and 'environmentalism' to illustrate what they mean. You can think of these generalized arguments as a special kind of what **Hinchliffe and Belshaw (2003)** call '**discourses**'. Hinchliffe and Belshaw use that term to describe practices which inform the way that people make sense of the world around them. One of the ways in which they identified such discourses as working was in the form of stories which made certain courses of action more thinkable or doable than others. Here the usage is very similar to that of the generalized arguments to which Thévenot et al. refer. We will encounter a number of **justifications** in the rest of this chapter, and you should get a better sense of what can count as a generalized argument or discourse in the context of GM food as we proceed.

discourse

justification

What evidence was offered to back up such a justification?

What should also become clear below is that just because a generalized argument or discourse is used to justify a given position, it does not follow that it will be accepted by those for whom it is intended. All justifications are very much attempted justifications in a sense, and it is because of this fact that the parties involved in a contest will invariably try to provide **evidence** of various sorts to further bolster their case. As we shall see in the case of GM, such evidence can take a variety of forms. Again, whether this evidence is treated as convincing or not is a different matter altogether.

evidence

What interests and resources were involved?

We have already noted that in addition to a justification and associated evidence, Thévenot et al. suggest that all public claims must be supported by what they call '**arrangements**' of various kinds if they are to stand a chance of being successful. Basically, you can think of such arrangements as corresponding to the assemblages of **resources** described by **Hinchliffe and Belshaw (2003)**. There it was suggested that the resources available to different interests will inevitably vary. Thus an environmental movement might bring together media coverage, protest, civil disobedience and lobbying, while a commercial interest may possess access to government and resources of wealth, investment and employment or the threat of withdrawal. Power considered as resources or arrangements, then, comprises those components that interests can bring together and deploy in order to exert leverage, influence or force. We will come across several examples when we consider the episodes of contest about GM food in Sections 4–6.

arrangement

resource

What was the outcome of the action?

The outcome of episodes of contest will obviously depend largely on how successfully the participants are able to justify, offer evidence for, and otherwise support their position (i.e. how they use discourses and assemble resources). We will note what happens in each of the three GM cases.

Summary

There are three requirements for making claims in the public arena:

- a generalized argument (or discourse) designed to justify one's position
- evidence of some sort to back up that argument
- an arrangement (or assemblage of resources) to support the case.

4   Episode 1: advertising GM

In most parts of the world at the turn of the twenty-first century the name of Monsanto became almost synonymous with biotechnology in general and genetic engineering or modification in particular. In part, as we shall see below, this was the result of a conscious decision made by the executives of the corporation in the late 1990s that it should become the principal proponent and cheerleader for the new techniques and their possibilities. However, Monsanto did not emerge from nowhere to play this role.

Monsanto was founded as a company in 1901 by a chemist in order to manufacture saccharin, the first artificial sweetener. It quickly diversified into the manufacture of industrial chemicals, and by the 1940s had become a significant supplier of plastics and synthetic fabrics. In the 1960s and 1970s it became infamous as a major producer of both PCBs and Agent Orange. The former was a chemical compound used for lubricants, coolants, and waterproofing but was discovered to cause cancers and subsequently banned. The latter was used by the US forces to defoliate the rainforests of south-east Asia during the Vietnam War. Exposure to toxins in it was subsequently proved to have caused skin rashes, pains in joints, weaknesses in muscles, neurological disorders and birth defects to some of those on the ground. Subsequently, the company slowly divested itself of its chemical interests and began dedicating its profits towards biotechnological research.

4.1 What was being contested?

In 1982, a Monsanto team led by Ernest Jaworsky, a cell biologist, succeeded for the first time in genetically modifying the cell of a petunia plant such that it and its offspring would be unaffected by the application of a particular antibiotic. Research developing and refining this and associated techniques progressed in two directions. The first was the development of crops genetically modified to be tolerant to the company's Roundup herbicide. This meant that, unlike the surrounding plants, they could be sprayed without being killed. The second was the engineering of plants such as Bt corn (discussed in **Bingham and Blackmore, 2003**), which would be resistant to various harmful insects and diseases. In 1995, the first US government approvals for the growing of Monsanto's genetically modified soya beans, potatoes and cotton were granted, and within three years more than 25,000 acres in that country were planted with soya beans resistant to Roundup.

Figure 1.5 'Frankenstein Food' headline, *Daily Mail*, 17 February 1999.

Awareness of the new kinds of crops quickly spread, but governments, regulators and consumers in other parts of the world – notably Europe – were not quite so keen to jump so early onto Monsanto's bandwagon. British public opinion in particular was hostile, with headlines alluding to the supposedly unnatural qualities of GM food (see Figure 1.5) and to Tony Blair's support for the technology (see Figure 1.6) already becoming commonplace on the front pages of British newspapers. Further, in March of 1998 the supermarket chain Iceland announced that they would not sell any products containing GM ingredients. In response, Monsanto's Chief Executive of that time, Robert Shapiro, approved a worldwide advertising blitz, including a million-pound campaign designed to win round the hearts and minds of the British people to the safety and benefits of GM crops. Over a three-month period, the company would place a series of full-page advertisements in the national weekend press (see Figures 1.7, 1.8 and 1.9). These were launched in the first week of June 1998.

The participants in the first episode of conflict over GM, then, were Monsanto on the one hand and significant elements within the British public on the other. And what was being contested was effectively **knowledge**. At the time there were no GM crops being commercially grown on British soil, and relatively few GM foodstuffs available to buy there. That was not important to Monsanto however. Their advertising campaign was not trying to sell products here and now, but to convince the British public that Monsanto knew best about GM food technology, that the benefits of what they were developing were substantial, and that they could be trusted as telling the truth about both of these assertions.

knowledge

Figure 1.6 'The Prime Monster', front page of *The Mirror*, 16 February 1999.

4.2 What justification was given for action?

I have suggested that what can be called generalized arguments or discourses play an important role in the way that groups or individuals defend their positions in environmental contests. The best place to start looking for these in this case is Monsanto's advertisements themselves.

Activity 1.3a

Read the first advertisement from Monsanto's campaign (Figure 1.7) and think a little about how the way it is phrased sits alongside what you now know Monsanto was trying to achieve with the campaign.

FOOD BIOTECHNOLOGY IS A MATTER OF OPINIONS. MONSANTO BELIEVES YOU SHOULD HEAR ALL OF THEM.

GENETICALLY modified food is the subject of much heated debate.

As a biotechnology company, at Monsanto we firmly believe in it. Of course, we're a business and aim for our shareholders to profit from this technology. However, our excitement and commitment to food biotechnology stems from the real benefits it provides for both consumers and the environment.

There are others with less supportive views. Some are openly hostile. It is only fair that you appreciate the spectrum of opinions before making an informed decision.

We're about to run an advertising campaign presenting the benefits of food biotechnology. As well as our views, we will be publishing the addresses and phone numbers of those with different views, including some of our most vocal critics, encouraging you to contact them. This may sound unusual, but we believe that food is so fundamently important, everyone should know all they want about it.

Beside the advertising, there are leaflets in many of your local supermarkets and you can call us free on 0800 092 0401 if you have questions or would like further literature. Alternatively visit our website at www.monsanto.co.uk.

Clearly our aim is to encourage a positive understanding of food biotechnology. And the truth is, we know it will take more than our words to convince you.

MONSANTO

To find out what others are saying, [contact] Iceland ... at www.iceland.co.uk Try contacting Friends of the Earth ... at www.foe.co.uk.

Figure 1.7 Monsanto advertisement: opinions on food biotechnology.

WORRYING ABOUT STARVING FUTURE GENERATIONS WON'T FEED THEM. FOOD BIOTECHNOLOGY WILL.

THE WORLD'S population is growing rapidly, adding the equivalent of a China to the globe every ten years. To feed these billion more mouths, we can try extending our farming land or squeezing greater harvests out of existing cultivation.

With the planet set to double in numbers around 2030, this heavy dependency on land can only become heavier. Soil erosion and mineral depletion will exhaust the ground. Lands such as rainforests will be forced into cultivation. Fertiliser, insecticide and herbicide use will increase globally.

At Monsanto, we now believe food biotechnology is a better way forward. Our biotech seeds have naturally occurring beneficial genes inserted into their genetic structure to produce, say, insect – or pest-resistant crops. The implications for the sustainable development of food production are massive: Less chemical use in farming, saving scarce resources. More productive yields. Disease-resistant crops. While we'd never claim to have solved world hunger at a stroke, biotechnology provides one means to feed the world more effectively. Of course, we are primarily a business. We aim to make profits, acknowledging that there are other views of biotechnology than ours. That said, 20 government regulatory agencies around the world have approved crops grown from our seeds as safe.

Food biotechnology is too great a subject to leave there. Ask for a leaflet at your local supermarket, write to us, call us free on 0800 092 0401 or visit our online Comments & Questions.

We're convinced biotechnology is a responsible way to provide food for the next century. We hope you agree.

MONSANTO

We urge you to hear all sides of the debate. For other views, contact Friends of the Earth on …at www.foe.co.uk. Visit the Food for Our Future website at www.foodfuture.org.uk.

Figure 1.8 Monsanto advertisement: the ability of food biotechnology to feed future generations.

Comment

Were you surprised by the language used here? On the face of it, this all seems eminently sensible and is quite different from a lot of advertisements one encounters where the fact that some huge corporation is desperately trying to convince us to buy their products is very transparent. On the other hand, perhaps this was precisely the point. Advertising, by its very nature, cannot force us to buy or even buy into something: it can only use persuasion. In that sense, we might see this first advertisement in Monsanto's campaign as cleverly designed to put us at our ease and establish some trust. See what you think after reading the second.

Activity 1.3b

Carefully read the second advertisement (Figure 1.8) and make a note of the environmental problems that Monsanto associates with overpopulation and how GM crops could reduce them.

Comment

In terms of environmental problems, I noted soil erosion, depletion of mineral reserves, deforestation and increased usage of chemical fertilizers and pesticides. Farming with GM crops, Monsanto argues, could be altogether more sustainable, reducing inputs while improving outputs.

progress

We are now in a much better position to understand how Monsanto justified its action in promoting GM. For Monsanto, GM crops offer improvement, or – to put it more starkly – GM is '**progress**'. As a discourse employed to defend a position, progress is a fairly safe bet. The notion has been an extraordinarily powerful one in western societies in particular, allowing individuals, groups, and nations to mark themselves out as better (or at least better off) than those who came before them. Relatively often, as in this case, the indicator of such progress is a scientific or technological artefact or process of some sort. Think, for example, of how we speak in terms of 'the industrial revolution', 'the atomic age', 'the television age' or 'the computer revolution'. Positioning GM food as part of a 'biotechnology revolution' was therefore potentially a very persuasive move by Monsanto. It is a very resonant one too as the discourse of progress appeals to some very deep-seated habits. Progress today, for example, is what makes us (or some of us at least) want to go out and buy the latest computer or DVD player or car or whatever.

Activity 1.3c

Having provisionally identified that Monsanto justified their actions in terms of the idea of GM as progress, read this last advertisement (Figure 1.9). Does its content lend any weight to our hypothesis?

IF IT WEREN'T FOR SCIENCE, HER LIFE EXPECTANCY WOULD BE 41 YEARS.

IN BRITAIN, reaching forty little more than a century ago was a major achievement. Vaccinations and developments in medicine have increased our life expectancy beyond our great grandparents' dreams.

But with each new scientific breakthrough, there is always concern. Is it safe? Is it ethical? Should it be trusted? Perhaps it is human nature always to distrust new scientific advances, just because they are new.

A recent example, plant biotechnology, is currently (and rightly) the object of such scrutiny. At Monsanto, we believe plant biotechnology produces better crops, like potatoes, soybeans and corn. These require less pesticide application, conserving scarce resources and reducing the cost to both the farmer and the environment.

Naturally, the produce of our technology is thoroughly tested for human and environmental safety. In fact, biotechnology crops undergo stringent safety testing where 'standard' ones are often not tested at all. So far, the government regulatory agencies of over 20 countries have approved Monsanto crops.

We hope you too will welcome biotechnology when you know the facts. Please ask for a leaflet at your local supermarket, write to us, call 0800 092 0401 or visit our online Comments & Questions form.

MONSANTO

We urge you not just to accept our word on plant biotechnology. Contact Greenpeace ... at www.greenpeace.org.uk, or Food for Our Future at www.foodfuture.org.uk.

Figure 1.9 Monsanto advertisement: the benefits of science.

Comment

For me at least, both the title and the text draw further on the notion of progress and so confirm our initial analysis.

4.3 What evidence was offered to back up such a justification?

By the very nature of advertisements of this kind, it is difficult to include much by way of 'proof' for the assertions made in them and Monsanto simply provide a website for further information. In the place of any other evidence, however, when it comes to ensuring that the claims are taken seriously there does seem to be a significant reliance on the authority of science and the principles of objectivity, neutrality and truth that have been traditionally associated with it. Did you notice the same thing? If so, how successful do you think that reliance was?

4.4 What interests and resources were involved?

Monsanto (in the first advertisement of the three above) were up front in the campaign, both about their commercial interests in the success of GM food – 'Of course, we're a business and aim for our shareholders to profit from this technology' – and their attempt to convince the British public that their interests coincided: 'our excitement and commitment to food biotechnology stems from the real benefits it provides for both consumers and the environment'. As for the resources assembled to support their position, the very fact that they were a multinational company which could afford to spend such a large sum of money advertising not yet existing products should give you a fair indication of the number of different entities that were linked together behind the scenes. Businesspeople, crops, laboratories, money, advertising agencies, genes, equipment and scientists were just some of the things involved in building an arrangement that looked at once solid, formidable, and difficult to stop.

4.5 What was the outcome of the action?

Despite this solid, formidable, and apparently difficult to stop set of supporting resources, despite the supposed authority of science and despite the powerful justifying discourse of progress, the Monsanto advertising campaign we have just reviewed was universally accepted to be a huge PR disaster – even by the firm itself. Three brief points should be enough to give a flavour of what happened. First, just days after the launch of the campaign, Prince Charles weighed into the debate on the anti-GM side, attracting a huge amount of media attention and thereby changing the profile and shape of the debate significantly. Second, a number of complaints made by environmental organizations (including the Royal Society for the Protection of Birds (RSPB), the Green Party and GeneWatch), as well as members of the public, concerning some of the stronger claims made in the advertisements were upheld by the Advertising Standards Agency. Finally, and most significantly, in a post-BSE climate in which trust in science-related institutions was already low, the majority of the British public were in general highly sceptical of the discourse of progress and the appeal to scientific rationality which Monsanto used to underpin the advertisements.

For more on trust see Bingham and Blackmore (2003).

Summary

- The episode of conflict over GM food between Monsanto and the British public centred on knowledge.
- Monsanto justified their position using a discourse of progress.
- Their claims were sustained using the authority of science.
- As a multinational corporation they were able to mobilize an extensive network of material resources to support their case and advance their interests.
- The advertising campaign which Monsanto ran was unsuccessful for a number of reasons.

5   Episode 2: trialling GM

During 1999, the British government set up the Farm-Scale Evaluation (FSE) programme in order to assess the wider effects on farmland wildlife of growing GM herbicide-tolerant crops in direct comparison with current (industrialized) farming practice. These trials – which were completed in 2003 – were covered by a voluntary agreement between government and an industry body known as SCIMAC (Supply Chain Initiative on Modified Agricultural Crops). The basis of that agreement was that biotechnology companies agreed not to grow GM crops commercially in the UK while the trials were taking place. The crops involved in the trials were oilseed rape, forage maize, sugar beet and fodder beet, and each GM variety was compared with an equivalent non-GM variety grown in the normal manner by the farmer. Research on the sites was carried out by a consortium of three independent contractors led by the Centre for Ecology and Hydrology, and monitored overall by the Independent Scientific Steering Committee which included representatives from English Nature, the RSPB and the Game Conservancy Council.

5.1 What was being contested?

In April 2001, a list of farms where that season's trials would be taking place was published by the government. One of the new sites was near a village called Wolston in the East Midlands (see Figure 1.10).

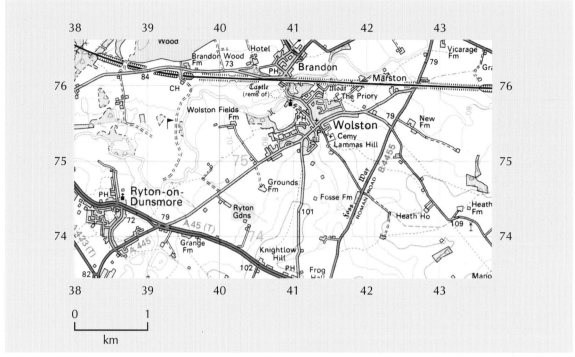

Figure 1.10 The location of the village of Wolston, Warwickshire.

Activity 1.4

The OS map reference provided by the government for the precise location of the trial in Wolston was given as 427 759. Using the gridlines marked on the map, find and mark on Figure 1.10 exactly where this would be.

You can find the answer to this activity at the end of this chapter.

Also near Wolston, in Ryton, is the headquarters of the Henry Doubleday Research Association or HDRA. Founded in 1954 as a club for experimenting gardeners, by 2001 HDRA had a membership of approximately 27,000 people interested in many aspects of organic horticulture and agriculture. Ryton Organic Gardens is an internationally famous tourist attraction where ten acres of display gardens and a large conservation area are managed entirely organically and where scientific work is carried out for the Soil Association (the body which certifies most of the organic food grown in the UK). As soon as the information about the proposed trial site in Wolston became publicly available, some very anxious noises started coming out of HDRA and a campaign began to have the trial halted.

Such is the background to the second episode of conflict over GM that we shall be analysing in this chapter. Unlike the first, which I described as being mainly about knowledge, what was contested in this case was mainly the practical **implications** for diversity and choice if GM crops became widely grown and

implications

sold in the UK – diversity and choice for farmers in terms of the kind of crops which they would be able to grow, and diversity and choice for consumers in terms of the kinds of foods they would be able to buy.

5.2 What justification was given for action?

Activity 1.5

Read the following briefing paper prepared by HDRA to explain their position on the trials (Figure 1.11 overleaf). This is a much longer document than the individual advertisements in the previous section and you will need to concentrate if you are to get something from it. In order to help guide your reading, try to find and note down brief answers to the following questions:

(a) If it went ahead, what would this particular trial have told us about the environmental consequences of GM crops?

(b) What was the main reason for HDRA's opposition to the trial?

(c) Are there any other negative environmental consequences of GM crops mentioned?

Comment

In answer to the three questions I posed, you should have got something like the following:

(a) The trial was designed to provide information about the effects on the local biodiversity of a variety of maize genetically modified to be resistant to a certain type of herbicide (which would allow farmers to spray whole fields with a particular brand of weedkiller without harming the crop).

(b) The principal reason HDRA opposed the trial is because of the risk of their organic crops being contaminated by pollen from the GM variety being tested.

(c) Other negative environmental consequences mentioned include the possible impact of GM on earthworms and soil micro-organisms, and the risk of transfer of the herbicide-tolerance gene to other organisms.

From the briefing paper, then, we learned that HDRA was concerned about the danger that GM crops posed to the organic status of Ryton Gardens. We also learned how they justified their position against the trials. Just as Monsanto drew on a discourse with wider applicability (progress) in order to argue for its specific claims, so too did HDRA. Their discourse can be summarized as one of '**purity**'. purity Although relatively new in its association with organic food, the notion of an unsullied or pristine nature has – like progress – a long history. Perhaps most notably it has been the basis for the valuing of wilderness, which you will learn more about in the next chapter. In the context of the GM controversy the associations of a purity discourse were certainly very powerful. Even by the time that this episode took place much of the negative coverage of GM food had been

HDRA Briefing Paper – GM Trials at Wolston

Trials Overview

Farm Scale Evaluations (FSEs) are being conducted throughout the UK on forage maize, spring and winter sown oilseed rape. All have been engineered to be tolerant to the herbicide glufosinate ammonium (Liberty). They have been developed by AgrEvo, which is part of a large multi-national, Aventis, formed when Hoechst merged with Rhone Poulenc to create a 'life science company', i.e. a pharmaceuticals/agrochemicals business. The trials are intended to run between 2000–2003 and, according to the DETR are concerned with biodiversity, not food or feed safety.

What will the trials be investigating?

Concern has been expressed about the possible impact of growing herbicide resistant crops on the environment and wildlife. If fewer weeds exist in farmers' fields, will the current decline in farmland birds get worse? Are mammals adversely affected? How are invertebrates affected? Do genes spread from GM crop plants to non-GM crop plants growing nearby, or to related weeds? …

The Fodder Maize trials

Fodder maize is grown specifically to be turned into silage to be fed to cattle. Twenty-five sites have been chosen in England, and three in Scotland, with the aim of having 32 sites in total. Each site will vary from 2–10 ha (5–25 acres) and will compare weeds and wildlife in the GM crop with a similar non-GM crop planted nearby. According to Genewatch, the Government pays all the research costs and the industry provides the seed, identifies the trial farms and pays farmers a 'compensation package', thought to be around £1,000 per hectare.

The maize, a variety called Chardon LL, has been genetically modified by incorporating a gene that makes it tolerant to the chemical glufosinate, a broad spectrum weedkiller, by incorporating a gene that detoxifies it. This will enable farmers to spray Liberty, killing all the weeds but leaving the crop alone.

Notification to HDRA

Even though maize will be sown from the end of April through May, the first thing HDRA knew about it was when we were rung up on 5 April for comment by the local BBC (Coventry and Warwickshire) radio station. According to a spokesman from Aventis, the company knew of the existence of the HDRA site but, as it considered that there was no risk from the GM crop, didn't bother to tell us. There is no statutory requirement for consultation.

Contamination risk from GM pollen

Maize is wind pollinated and, whilst most pollen falls within a few hundred metres, small quantities of pollen are likely to travel much further under suitable atmospheric conditions, according to the National Pollen Research Unit. Each maize plant can produce up to 50 million pollen grains and pollen can remain viable from between 3 hours and 9 days. Bees could also move maize pollen several miles from the crop each day.

The possible effect of the trial on HDRA's organic status

The public expects organic food to be free from contamination caused by genetic engineering, hence the decision of the Soil Association, the UK's largest organic accreditation body, to lay down strict rules to prevent this from happening. Land at Ryton Organic Gardens is accredited by the Soil Association, which enables fruit and vegetables grown there to be sold as organic in the shop and restaurant on site. A separate trials area is used for research to develop improved organic methods. This work is aimed at commercial horticulture, and most of the trials are funded by the Ministry of Agriculture (MAFF).

△ top

HDRA Briefing Paper – GM Trials at Wolston

Cont'd

Because of the risk of GM maize pollen contaminating the Ryton site and cross pollinating with the sweetcorn growing there, HDRA is at risk from losing its organic certification. Soil Association standards, which are legally binding, state that 'organic certification may be withdrawn from land, crops, or products where … the Certification Committee considers there is contamination, or specific risk of contamination from GMOs or their derivatives'. Should that happen, HDRA would be forbidden from selling its own produce, carrying out trials, or even, possibly, from calling itself an organic centre, for a period of at least two years.

Shortcomings of the trials

HDRA is calling for the Wolston trial to be abandoned because of the unacceptable risk to its organic status. However, like many other environmental organisations, we have serious misgivings about the scientific methodology of the trials. In the short time allotted to the trial, any subtle environmental changes are unlikely to show up, nor is there to be any monitoring of earthworms and soil micro-organisms which are crucial to maintaining soil fertility.

We also believe that field trialling is premature, given the question marks that still hang over genetic engineering technology. There is preliminary evidence, for example, that the herbicide resistant gene incorporated into GM maize is able to transfer to bacteria living in the guts of honey bees. Could a similar transfer occur in cattle fed silage made from genetically engineered maize? The whole licensing process of genetically engineered crops is proceeding far too quickly and is not in the public interest.

Scientists claim that Chardon LL maize has been rigorously tested. In fact, there is no evidence that it has been tested on cattle, for which the forage is intended, and there are several other flaws in the tests that have been carried out.

12 April 2001

△ top

Figure 1.11 Extracts from HDRA Briefing Paper – GM Trials at Wolston.

Figure 1.12 'Keep
Nature Natural' fish-
strawberry logo –
negative coverage of
GM food.

premised on the fact that it was unnatural or impure. The imagery of half-fish, half-fruit beings (see Figure 1.12) or the language of 'Frankenstein Foods' (see Figure 1.5) was ubiquitous. Organic food, on the other hand, had a very different image – one of being untainted by modern science and technology.

5.3 What evidence was offered to back up such a justification?

Although critical of what it described as its 'unsound' versions, HDRA (like Monsanto before it) drew on science to sustain its case. Whereas Monsanto appealed to the values traditionally associated with science, however, HDRA referred to specific research carried out by the National Pollen Research Unit, which it felt assisted its argument that GM trials would threaten the purity of Ryton's crops. Since the alleged possibility of cross-pollination between the maize being grown on the test site and sweetcorn being grown at Ryton Gardens was crucial to HDRA's position, it is worth exploring this aspect in a little more detail. (Box 1.2 explains the process of pollination.)

pollination

anther

stigma

Box 1.2 Pollination

Pollination is part of the process of the sexual reproduction of plants, specifically the transfer of pollen from the **anther**, where it was made, to a **stigma** (see Figure 1.13). In some species of plants, such as the garden pea, it is usual for pollen to be transferred to the stigma of the same flower: this is called self-pollination. Many other species – such as corn – have mechanisms which ensure that the pollen is transferred from a different individual of the same species. As neither plants nor grains can move actively from place to place, other agents are used to transfer the pollen grains. These include animals (such as insects, birds, bats and mice), wind and – in a few cases – water currents.

The crucial question in this case is how far maize pollen travels in the process of cross-pollination. According to the government guidelines – to which the trial site at Wolston would have to keep – if the distance between the GM maize and the organic sweetcorn was 200 metres, then the likely frequency of cross-pollination would result in no more than one sweetcorn kernel in every 40,000 being a GM hybrid. This was deemed to be an acceptable separation and indeed became the standard for GM trials. However, such an assessment was disputed, and according to a report commissioned by the Soil Association, carried out by the National Pollen Research Unit (NPRU) and referred to in the HDRA briefing paper, the transport of maize pollen over distances greater than 200 metres is far more common. In fact, although maize pollen is relatively large and heavy (and thus its dispersal tends to decline relatively rapidly with distance), taking into account a range of possible wind speeds and the fact that bees can transport pollen grains several miles within the 24 hours or so within which they are still viable, the NPRU conclusion was that in moderate conditions the cross-pollination rate would be more like one kernel in 93. Acting on such information,

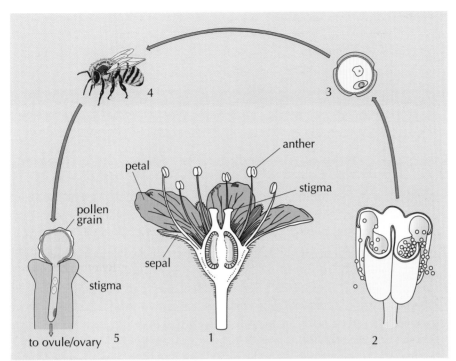

Figure 1.13 Pollination: the route taken by the pollen and the role of pollinating insects. (1) The flower with its anther, stigma and visually striking petals and sepals. The ripe anther (2, in cross-section) releases the pollen (3). An insect (4) transports it to the stigma of another flower of the same species. Having arrived on the correct stigma (5), the pollen grain germinates; the pollen tube grows down to the ovary, where the ovum is produced and fertilized.

the Soil Association recommended that there should exist a 'notification zone' of six miles around organic farms such that if GM production was planned within that area then a decision on the risk of contamination could be made.

Activity 1.6

Returning to Figure 1.10, calculate the distance between the location of the trial site you marked in Activity 1.4 and Ryton Gardens which is already indicated on the map. How does this distance compare with the figures of 200 metres and 6 miles (approximately 10 km) that have just been discussed?

You can find the answer to this question at the end of this chapter.

It is perhaps worth noting that we are not just dealing with theoretical claims here. In 1998, some 87,000 packs of tortilla chips worth over £100,000 were recalled and destroyed in the UK after a routine analysis revealed that transgenic DNA was present in the product. The maize used had been grown on a 7,000 acre organic farm in Texas, a region in which many farmers grow GM maize. After an investigation, the suppliers concluded that the most likely source of contamination was cross-pollination.

Having learned a little about the background, do you think that HDRA were right to be concerned about contamination? Who should be held responsible if cross-pollination from a GM plant means that an organic crop or product has to be destroyed, potentially costing a farmer or company many thousands of pounds?

5.4 What interests and resources were involved?

Like Monsanto, HDRA obviously had interests that they were seeking to advance during this episode of contest, both in the immediate sense of preserving the organic status of their headquarters and in the broader sense of promoting organic farming practices and food consumption. Unlike Monsanto, however, they had few human, technical or financial resources to attach to those interests in order to support their case. This meant that they had to adopt a different strategy and work hard at building coalitions with other groups and individuals whose interests they could align with their own. These included (in different ways) the Soil Association with its direct connections to HDRA, local residents who knew the gardens and who also may have had their own reasons for not wanting a GM trial in the vicinity, environmental campaigners at a more national level who had been taking direct action against previous GM trial sites (see Figure 1.14), and the media in various forms for whom this was simply a great news story. However, even HDRA could never appear to the outside world as formidable or as solid or in any way as difficult to resist as a Monsanto, simply

Figure 1.14 Protestors destroying GM crops at a UK trial site.

because those alliances were provisional and temporary (for more on coalition building and alliance making see Chapter Three). That did not stop an effective support for HDRA's position being built, however. Shortly after the trials were announced (the timing was crucial as it was only weeks away by that stage) a great deal of lobbying, protesting, awareness raising and reporting took place at local, regional and national levels.

5.5   What was the outcome of the action?

If we ask whether the combination of justification, proof and material support that HDRA employed in the episode was successful then the answer has to be yes on two counts. In the first place, the pressure orchestrated by HDRA resulted in the cancellation of the trial at Wolston. After the conservation director of the RSPB threatened to withdraw his organization from the scientific steering committee, the government minister responsible for the trials programme asked SCIMAC to abandon their plans. They agreed on the condition that it was made clear that their decision was in the interests of public confidence and not a reflection of any scientific or legal basis not to proceed. In the short term this decision applied only to Ryton and did not signal any general change of policy regarding trials near organic farms. However, in the longer term the sorts of arguments made by HDRA did come to fruition in a second way when, in response to a report on the matter, the government admitted in early 2002 that there were legitimate concerns about the way in which the FSE programme had operated.

Summary

- The episode of conflict around GM food between HDRA and the trial programme centred on implications for diversity and choice.
- HDRA justified their position using a discourse of purity.
- Their claims were sustained using scientific evidence related to cross-pollination.
- As a small organization they had to build coalitions to support their case and advance their interests.
- The campaign against the GM trial at Ryton was successful.

6   Episode 3: resisting GM

Karnataka is a state of India, situated on the south-west coast (see Figure 1.15). The capital is Bangalore. Agriculture has traditionally been strong in this area and remains the occupation of about 80 per cent of the population, but recently the state has exploited its position as one of the most educationally advanced in India to diversify into the information technology sector. At the end of the twentieth century, Karnataka accounted for approximately 30 per cent of software from India, with about 90 per cent of that amount – produced mainly in the global IT centre of Bangalore – exported to the USA. However, after the economic slowdown in the USA led to diminishing trade and job prospects in IT, the Karnataka state government turned its attention to attracting still booming biotechnology industries. Tax concessions, low-tariff energy and floor space for companies involved in drug research, bioinformatics and genetic engineering were all advertised. Monsanto and their GM cotton variety Bollgard were a particular target for a state which produces the fourth largest amount of cotton in the country.

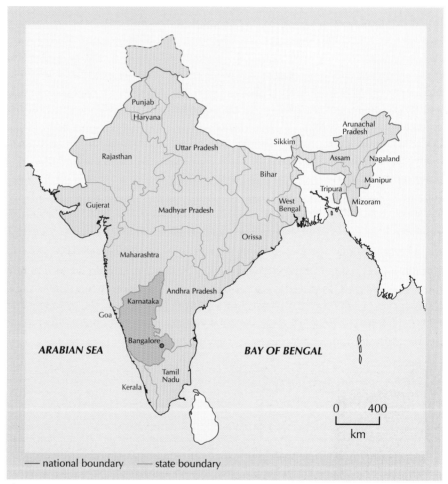

Figure 1.15 India, showing the location of Karnataka and Bangalore.

6.1 What was being contested?

At the same time that the process of state encouragement was taking place, biotechnology research and biotechnology corporations were increasingly becoming a target of the Karnataka Raiya Ryota Sanghe (KRRS) or Karnataka Farmers Association. The KRRS was founded in 1980 in order to tackle popular grievances like debts, agricultural prices and discriminatory taxes. However, it soon started to develop a broader vision, loosely based on the teachings of

Mahatma Gandhi, the objective of which was the realization of a 'village republic'. This was a form of social, political and economic organization based on direct democracy, economic and political autonomy and self-reliance, and the participation of all members of the community in decision making about their common concerns. During the 1980s and 1990s, these ideals led members of the KRRS to take non-violent direct action against the property of multinational companies (Cargill in 1993, KFC in 1995) who represented what they saw as a very different future. By 1998 it was the turn of Monsanto.

Not long before, Monsanto had identified India as one of the main global markets for its GM plants, and had set about trying to make that vision a reality. It paid US $20 million for the country's most advanced genetic engineering laboratories and spent an estimated US$4 billion purchasing companies that were leading Indian seed suppliers or had access to that market. Monsanto's great hope for breaking into India was its brand of cotton – the aforementioned Bollgard – which was genetically modified to contain the Bt gene (see **Bingham and Blackmore, 2003**). For the cotton, the gene does a similar job in protecting against the bollworm – a major cotton pest worldwide – as it does for the Bt corn against corn borer moth larvae. However, in November 1998 the leader of the KRRS, Professor Nanjundaswamy (see Figure 1.16), and a group of farmers arrived at two of Monsanto's Bt cotton test sites and, after first paying the farmer compensation, cut down the cotton crops and set fire to them (see Figure 1.17).

If it was knowledge that was being contested in the first episode we examined, and the implications for diversity and choice in the second, then we find here a third aspect of the GM issue taking centre stage. For what was being contested in Karnataka first and foremost, as we shall see, were the effects of the production of GM foodstuffs for people's (and especially small farmers') control over their **livelihoods** (for more on livelihoods see **Castree, 2003**).   livelihood

Figure 1.16 Professor Nanjundaswamy, leader of the KRRS.

Figure 1.17 Farmers burning Bt cotton, Karnataka State, India.

6.2 What justification was given for action?

Activity 1.8

After studying the previous two episodes, you should now be starting to get some idea of what can count as the sort of discourse which can be appealed to in order to defend a position. To test this out read through the announcement (reproduced here as Figure 1.18) – circulated by email by Professor Nanjundaswamy— and try to form an opinion of how the action of the KRRS in burning the GM crops was justified. Remember, you are looking for some sort of argument, story or myth, which applied to the situation but also had a set of wider resonances. If the actions of Monsanto in the first and the HDRA in the second episodes were justified by 'progress' and 'purity' respectively, what is the justification for action here?

Comment

The first thing that is likely to have struck you while reading through this piece was how different it was in style to the documents which we looked at in the previous two episodes. Whereas both the persuasive rhetoric employed by Monsanto in their advertising campaign and the sober measured language of the HDRA briefing paper seemed designed to convince the reader of something, Professor Nanjundaswamy's email release feels more like a call to (non-violent) arms.

As to how far that call to arms was justified, it seemed to me at least that – while not often explicit – underlying the statement was a discourse of **self-determination and tradition**. GM food was described as a manifestation of 'neoliberal globalisation', 'an instrument of capitalist domination', employed by 'corporate killers' and thereby a threat to ordinary Indians and their ways of

self-determination and tradition

```
Date: Thu, 26 Nov 1998 17:48:27 +0530
To: (Recipient list suppressed)
From: "PROF. NANJUNDASWAMY"
Subject: Cremate Monsanto!!
Mime-Version: 1.0
Content-Type: text/plain; charset="iso-8859-1"
Content-Transfer-Encoding: quoted-printable
```

=A0 ** please distribute as widely as possible **

Dear friends,
Monsanto's field trials in Karnataka will be reduced to
ashes, starting on Saturday. Two days ago the Minister
of Agriculture of Karnataka gave a press conference where
he was forced by the journalists to disclose the three
sites where field trials with Bt cotton are being conducted.
KRRS activists have already contacted the owners of these
fields, to explain to them which action will be taken,
and for what reasons, and to let them know that the KRRS
will cover any losses they will suffer. On Saturday the
28th of November, at midday, thousands of farmers will
occupy and burn down the three fields in front of the
cameras, in an open, announced action of civil disobedience.

These actions will start a campaign of direct action by
farmers against biotechnology, called Operation 'Cremation
Monsanto', which will not stop until all the corporate
killers like Monsanto, Novartis, Pioneer etc leave the
country. Farmers' leaders from the states of Maharastra,
Gujarat and Madhya Pradesh (states where Monsanto is also
conducting field trials) were yesterday in Bangalore to
prepare the campaign.

The campaign will run under the following slogans:

- STOP GENETIC ENGINEERING
- NO PATENTS ON LIFE
- CREMATE MONSANTO
- BURY THE WTO

along with a more specific message for all those who have
invested on Monsanto:
YOU SHOULD RATHER TAKE YOUR MONEY OUT BEFORE WE REDUCE
IT TO ASHES.
We know that stopping biotechnology in India will not be
of much help to us if it continues in other countries,
since the threats that it poses do not stop at the borders.
We also think that the kind of actions that will be going
on in India have the potential not only to kick those
corporate killers out of our country: if we play our
cards right at global level and coordinate our work, these
actions can also pose a major challenge to the survival
of these corporations in the stock markets. Who wants to
invest in a mountain of ashes, in offices that are
constantly being squatted (and if necessary even destroyed)
by activists?

The email from Professor Nanjundaswamy (Figure 1.18) continues overleaf.

For these reasons, we are making an international call
for direct action against Monsanto and the rest of the
biotech gang. This call for action will hopefully inspire
all the people who are already doing a brilliant work
against biotech, and many others who so far have not been
very active on the issue, to join hands in a quick,
effective worldwide effort.

This is a very good moment to target Monsanto, since it
has run out of cash in its megalomaniac attempt to
monopolise the life industry in record time. It is going
now through a hard time of layoffs and restructuration
in a desperate effort to survive, since it cannot pay its
bills. It is also a good time because several recent
scandals (like the pulping of the Monsanto edition of The
Ecologist, the whole "Terminator Technology" affaire, the
illegal introduction of Bt Cotton in Zimbabwe, etc) have
contributed to its profile as corporate killer, which,
being the creators of Vietnam War's Agent Orange and rBHG,
was already good enough, anyhow.

We are hence making a call to:

•   Take direct actions against biotech TNCs, particularly
Monsanto (be it squatting or burning their fields,
squatting or destroying their offices, etc)
•   Maintain the local or/and national press informed
about all the actions going on around the world
•   Take direct actions at stock exchanges targeting
Monsanto, to draw attention on its state of bankruptcy

We are making this call for action on the line of Peoples'
Global Action (PGA), a worldwide network of peoples'
movements, in order to emphasize the political analysis
beyond our opposition to biotechnology. This analysis
does not only take environmental concerns into account,
and is not limited to the defence of food security – it
attacks neoliberal globalisation as a whole, the WTO
regime as its most important tool, and the global power
structures (G8, NATO, etc) as the root of all these
problems. You will find the complete political analysis
on the manifesto of the PGA.
The fact that this call for action takes place on line
with PGA also has other implications:

•   We are calling ONLY for non-violent direct actions.
Non-violence in this context means that we should respect
all (non-genetically modified) living beings, including
policemen and the people who work for these TNCs
•   The campaign will take place in a decentralised
manner, and nobody should speak on behalf of other people
involved in the campaign without their consent (also not
on behalf of PGA, of course); however, people are welcome
to report about the actions of others without pretending
to represent them.

Friendly greetings,

Prof. Nanjundaswamy
President, Karnataka State Farmers Association

Figure 1.18 'Cremate Monsanto!!': Prof. Nanjundaswamy, email, November 1998.

life. The only solution is to 'kick [them] out of our country'. Some of the relationships among capitalism, environments and globalization are dealt with in **Castree (2003)** and you will learn more about anti-globalization protests in Chapter Five. However, Professor Nanjundaswamy's description elsewhere of western capitalism and especially 'the World Trade Organisation, the World Bank, the European Round Table of Industrialists, the World Economic Forum, the International Chamber of Commerce, the IMF, the G8, [and] the Bretton Woods organisations' as 'rotten, corrupt, and plundering the poor' (Vidal, 1999) probably gives you enough of the picture for now. And you will not need me to tell you about how effective arguing for the need to defend one's home, locality or country against outsiders can be in all areas of life.

6.3   What evidence was offered to back up such a justification?

Like the discourse or justification, what was used to back up the argument was largely implicit in the released statement. In this case it was because much of what counted as 'evidence' was assumed to be already possessed by the readers themselves in the form of memory or general knowledge – much of the case against GM food made by the KRRS was based on the experiences of a previous occasion in which agricultural 'progress' had been imported from outside: the so-called '**Green Revolution**'.

Green Revolution

The process that created the Green Revolution began in the mid-twentieth century when attempts were being made to repeat the success in tropical Asia that farmers in China, Taiwan and Korea were having growing new varieties of the japonica subspecies of rice. Such efforts focused on a search for fertile hybrids between these new japonica varieties and the indica subspecies more commonly grown in tropical Asia. The establishment in 1960 of the International Rice Institute (IRRI) in the Philippines with funding from the US-based Ford and Rockefeller foundations provided a focus for this work. Then, in 1962, the Institute used breeding techniques to produce a short-strawed variety called IR8 with grain type intermediate between japonica and indica. In the following two decades, many (so-called) high-yielding varieties were produced at the IRRI and by national research organizations, and were introduced throughout the rice-growing areas of the Indian subcontinent, Indo-China, Malaysia, Indonesia and the Philippines (see Figure 1.19). As a result, rice production in these areas increased by 60 per cent and countries which traditionally imported rice were able to achieve a large measure of self-sufficiency, or even become exporters of rice. It was this transformation that became known as the Green Revolution.

On the face of it, this sounds like a success story in which a technical development resulted in a significant improvement in the living conditions of many millions of people. And indeed, there is little doubt that (in the short term at least) the Green Revolution represented a formidable achievement, which was responsible for feeding many individuals who would otherwise have probably starved. However, over time it emerged that the introduction of the new seeds may have had a number of other consequences aside from producing high yields.

According to several commentators, these included some or all of the following: an unequal distribution of benefits between different social classes and regions (because the seeds could only grow under certain conditions); losses of biodiversity, increased soil erosion, contamination of ground water, eutrophication of water bodies and depletion of water tables (due to the large inputs of irrigation, chemical fertilizers and pesticides required by the seeds); a loss of traditional subsistence and seed-sharing practices (because of the new farming methods necessary to grow the seeds); and a failure on the part of the state to address pressing social issues such as land reform (as, encouraged by the

Figure 1.19 Peasants transplanting rice shoots in paddy field.

fertilizer and pesticide lobbies, the government became interested in only a certain version of agricultural modernization).

You can probably see why, if this version of the Green Revolution was prevalent among certain sectors of the Indian agricultural population (such as those represented by the KRRS), Professor Nanjundaswamy exploited it when campaigning against what was widely described by GM proponents as a 'second Green Revolution'.

6.4 What interests and resources were involved?

As a movement dedicated to the creation of a certain model of social organization, the interests of the KRRS in confronting the introduction of GM food (which they saw as a symbol of a very different future) to India are quite clear. It is also clear that the kinds of resources that need to be assembled in order to destroy crops and burn a field are not that great. However, that conflagration was only the symbolic first blow in a wider campaign by the KRRS against Monsanto's India venture. Locally, the organization's solidity was fairly robust, but Nanjundaswamy wanted to use this event to extend their network of influence much further. In order to do this more resources needed to be attached to their interests: specifically technological resources. By posting the statement that you read above – documenting and justifying the actions taken by the KRRS – to Internet mailing lists and websites, Nanjundaswamy sought to establish alliances with individuals and groups elsewhere whose interests coincided with his own. For this act if nothing else Operation 'Cremation Monsanto' attracted worldwide attention.

6.5 What was the outcome of the action?

Initially the campaign against GM food in India waged by the KRRS and other organizations was partially successful in its aims. The pressure created was certainly influential in the July 2001 decision of the Indian national government to extend (much against the wishes of the Karnataka state authorities) the testing period of GM crops and thus delayed commercial planting. However, by early 2002 and after counter pressures from the biotechnology industry, the Indian government finally gave the green light for full-scale cultivation. No one involved thought that the conflict would end there though, and you might like to try to find out what the state of play is when you are reading this.

Summary

- The episode of conflict over GM food between the KRRS and Monsanto centred on livelihoods.
- The KRRS justified their position using a discourse of self-determination and tradition.
- Their claims were backed up by appealing to a certain version of the consequences of the Green Revolution.

- As an existing political organization, the KRRS were able to mobilize a variety of resources locally to support their case and advance their interests, but also attempted to expand their sphere of influence during the conflict.

- Although the campaign against GM in India met with some initial success, it was ultimately unsuccessful.

7   Conclusion

You have encountered a wide range of material in this chapter, from detailed descriptions of scientific processes to invitations to join non-violent direct action against multinational companies. That diversity reflects in some small way the breadth of issues involved in what became the controversy surrounding GM food. To conclude, we will now return to our three chapter questions in order to review how what we have learned helps us to answer the task we set ourselves.

Why has GM food become so controversial? To address this first question we looked (in Section 2) at how the necessary ingredients for contest over GM (a shared object of attention and divergent assessment of its consequences) were present in the late 1990s and how the then latent contest became a conflict (characterized by visibility, the deployment of power to advance interests, commitments and polarization) at the turn of the century. We also saw in Sections 4, 5, and 6 how the number of different sub-issues involved (leading to conflicts over knowledge, implications and control) served to make GM increasingly controversial.

How has the operation of power enabled and constrained action in the GM food controversy? To address this second question we introduced (in Section 3) some requirements for making claims in the public arena, which enabled us to focus (in Sections 4, 5 and 6) on how actions taken by participants in three episodes of conflict about GM food were justified and supported. This allowed us to develop two approaches to power – as resource and as discourse – and to formulate a number of questions to follow how power operated in each of the episodes. The questions were these. What was being contested? What justification was given for action? What evidence was offered to back up such justifications? What interests and resources were involved? What was the outcome of the action? You might like to keep these in mind as you meet other examples of contest and conflict both in the rest of the book and elsewhere.

Can paying attention to the operation of power offer us ways out of the GM controversy? I have left addressing this final question until now because instead of adding to the substantive content of the chapter I wanted to leave you with an example of how the concept of power can be used to think through an environmental situation. In short, my answer is 'yes' – paying attention to the operation of power can offer us ways out of the GM controversy – for two reasons. First, it demonstrates the problems with letting contests become conflicts. If you recall, early in the chapter I referred to a report by a group of academics which predicted that the latent contest in the UK over GM food would soon become a full-blown conflict. One of the suggestions that they proposed for reducing this possibility was to develop ways of tracking and understanding how the different parties involved made sense of the GM issue. In the absence of such advice being taken seriously, what we ended up with was a series of completely polarized clashes, not only in the UK but in other countries as well. In each of these conflicts, complex situations were simplified and GM food was caricatured. In the first example we looked at it in this chapter it was caricatured as 'just' technical progress, a carrier of guaranteed benefits for all. In the second it was caricatured as 'just' impure, a threat to supposedly more 'natural' organic produce. In the third, it was caricatured as 'just' an expression of capitalism, a destroyer of livelihoods. You should feel confident to use the issue of power as a means of being critical of all claims such as these. They hardly sit well, for example, with our analysis at the beginning of the chapter that food issues are always characterized by an entanglement of scientific, technical, and social factors and that their proper solution requires similarly interdisciplinary and fine-tuned approaches which recognize that entanglement.

Fortunately, there are individuals and groups – including many of the individuals involved in the 1995 report (ESRC, 1999) – committed to inventing new ways of representing issues like GM, ways that do justice to their complexity and do not rely on caricature. If you are interested you might like to follow up in particular the work that the charity Action Aid (1999) has done with what are called 'citizens juries' in India, and the results of a technique called multi-criteria mapping developed by Stirling and Mayer (2001). Both attempt to make visible the different points of view that exist on GM food without reducing them to one or other side of a divide.

The second reason why I would suggest that paying attention to power can help point a way out of the GM controversy is that it can help us to decide where our sympathies should lie. However critical one might be of the sorts of discourses and justificatory strategies that have been employed in the episodes which we have reviewed in this chapter, there are occasions when one has to take a stand on issues like GM food even if only temporarily and non-definitively. This might be because you have or want to become involved directly in the debate, or might be because you find yourself in a seemingly banal situation – as when I as a vegetarian have to decide in a supermarket whether I should buy the cheese with cow rennet or the alternative which has a GM but non-animal substitute.

For my own part I have found that adding a final question to the five we asked of the protagonists in the episodes described earlier is useful, and that is: who benefits? This is not quite the same as asking what interests are involved because it draws more attention to the relationships between groups. The simple fact of the matter is that not all the interests of all those involved in the GM controversy can be satisfied. Resources of power are unequally distributed and we have seen plenty of examples of that in this chapter (see also: **Blowers and Smith, 2003; Hinchliffe and Belshaw, 2003**). My worry is that if production of GM foodstuffs increases within the current (2002) framework it will be the biotechnology corporations that will benefit disproportionally. Moreover I worry that they will benefit at the expense of the less powerful individuals and groups involved (both human and non-human) whether they are organic farmers, Indian smallholders or whatever, and that existing inequalities may be further exacerbated as a result. In turn this leads me to sympathize much of the time with those with least resources because they have most to lose. This is far from the same thing as being opposed in principle to the manipulation of life at a genetic level or the possible outcomes of such manipulation, like GM foodstuffs. Rather, it is just my own attempt to take seriously both the multiple possible consequences – intended and unintended – of the introduction of new scientific and technological innovations, and how much food matters.

You are obviously quite at liberty to disagree with my assessment of the situation (this is a book on contested environments after all), but by showing you some of my own thoughts on power, action and GM, I hope to have left you an initial sense of the importance of paying attention to environmental inequalities, a theme which will be developed further throughout the book.

References

Action Aid (1999) *Action Aid Citizens Jury Initiative: Indian Farmers Judge GM*, London, Action Aid.

Bingham, N. and Blackmore, R. (2003) 'Who knows? How risk and uncertainty affect environmental responses' in Hinchliffe, S.J. et al. (eds).

Blowers, A.T. and Smith, S.G. (2003) 'Introducing environmental issues: the environment of an estuary' in Hinchliffe, S.J. et al. (eds).

Castree, N.C. (2003) 'Uneven development, globalization and environmental change' in Morris, R.M. et al. (eds).

ESRC (1999) *The Politics of GM Food: Risk, Science, and Public Trust*, Special Briefing No.5, University of Sussex.

Fernández-Armesto, F. (2001) *Food: A History*, London, Macmillan.

Grove-White, R., MacNaughten, P., Mayer, S. and Wynne, B. (1995) *Uncertain World: Genetically Modified Organisms, Food and Public Attitudes in Britain*, Lancaster, CSEC.

Hinchliffe, S.J. and Belshaw, C.D. (2003) 'Who cares? Values, power and action in environmental contests' in Hinchliffe, S.J. et al. (eds).

Hinchliffe, S.J., Blowers, A.T. and Freeland, J.R. (2003) (eds) *Understanding Environmental Issues*, Chichester, John Wiley & Sons/The Open University (Book 1 in this series).

Morris, R.M. (2003) 'Changing land' in Morris, R.M. et al. (eds).

Morris, R.M., Freeland, J.R., Hinchliffe, S.J. and Smith, S.G. (2003) (eds) *Changing Environments,* **London, John Wiley & Sons/The Open University (Book 2 in this series).**

Stirling, A. and Mayer, S. (2001) *Rethinking Risk,* SPRU Report, University of Sussex.

Thévenot, L., Moody, M. and Lafaye, C. (2000) 'Forms of valuing nature: arguments and modes of justification in French and American environmental disputes' in Lamont, M. and Thévenot, L. (eds) *Rethinking Comparative Cultural Sociology,* London, Cambridge University Press.

Vidal, J. (1999) *Guardian Weekend,* June 19.

Whatmore, S. and Thorne, L. (1997) 'Nourishing networks: alternative geographies of food' in Goodman, M. and Watts, D. (eds) *Globalising Food,* London, Routledge.

Answers to activities

Activity 1.2

(a)   the bacterium Bt

(b)   corn

(c)   produces a protein that is toxic to certain insects

Activity 1.4

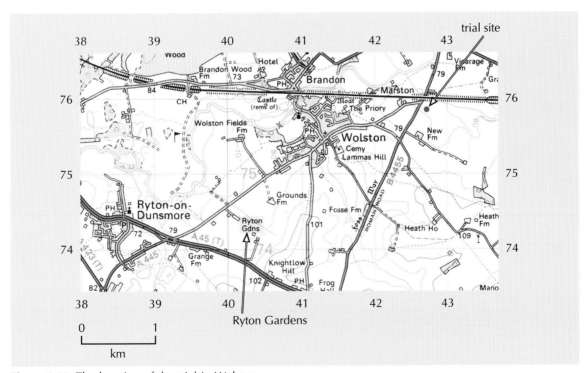

Figure 1.20 The location of the trial in Wolston.

Activity 1.6

On the map the distance between Ryton Gardens and the GM test site is approximately 6 cm, making the actual distance approximately 3 km or 3,000 m.

Chapter Two

Landscape, parks, wilderness

Chris Belshaw

Contents

1   Introduction

There are two main aims in this chapter. First, I want to explore some of the relationships between certain different parts, and differently thought of parts, of the surface of the world. Landscapes, parks and wilderness areas are fairly clearly linked in some ways. But we can consider these notions, and the links between them, in more detail. And, in particular, we'll use one of the analytical concepts of the book and consider certain of the different ways in which these environments can be thought to have *value*. Instrumental value – valuing something as a means to an end – is relatively uncomplicated. Non-instrumental value – valuing something for its own sake – is more complicated. Our focus here will be on value of the latter kind as understanding its complexity is critical if we want to answer one of the book's key questions 'how are environmental understandings produced?'

The other aim of the chapter follows from the first and speaks to one of the book's other key questions 'how and why do contests over environmental issues arise?' The ways in which we value land and landscape very much determine what we want to do with it. But because different people have different values, and want different things, the ways in which land is used and managed are, almost inevitably, sources of conflict and dispute. The second aim, then, is to explore the ways in which people disagree about the values, uses and future of landscapes, parks and wilderness, to consider how these contests arise and to consider some of the ways in which they might be resolved.

These two aims are, of course, closely connected. There is little point in discussing value just in the abstract, without considering how those values impinge upon the environment. Likewise, we can't say very much about how we manage, use and think about land without exploring these value questions in some detail.

In this chapter you will:

- learn how notions of aesthetic value, spiritual value and intrinsic value bear upon our understandings of landscapes, parks and wilderness;
- discover how our understanding and appreciation of the natural world is not simply a given, but is subject to historical and cultural influence;
- explore some of the ways in which conflicting values impinge upon both the development and management of national parks and the designation of wilderness areas;
- learn about deep ecology, and about challenges to deep ecological thinking that have been derived from both within and without the western tradition of environmental thought.

In tackling these issues we will engage in several ways with the theme of this book. One of our main focuses is on very large tracts of undeveloped or underdeveloped countryside. There is a complex history to our understanding of and relation to this land. And, inevitably, there is also much conflict about how such land should be used, or indeed whether it should be used at all. We need to consider how and why such conflicts arise, and the factors that bear on their resolution. Inevitably, again, the unequal power relations holding between parties to such conflicts help determine their outcome. So we need also to consider and reveal something of the agendas, often hidden, of those engaged in such conflict.

2   Landscape

We begin with the sort of scene that for many is reassuringly familiar: the sparsely-populated expanses, with the lakes, mountains, hills, found in the Scottish Highlands, Snowdonia, the Lake District, the west of Ireland, and many other places. Familiar because, even for those who don't regularly visit such places, landscapes like these figure so often in paintings, films, advertisements, and so on. And reassuring because such landscapes are places we feel we understand: they're never much more than an hour or so from a city, we can find maps and tourist centres and the wildlife is no longer to be feared.

Figures 2.1 to 2.3 give examples of such landscapes, all of which are the kinds of places that lots of people think are beautiful and should be protected, not only for our enjoyment, but also for that of future generations, and perhaps, too, simply for their own sake. Such landscapes, many of us believe, are highly valuable.

Figure 2.1 *Loch Coruisk and the Cuchullin Mountains, Isle of Skye*, c.1828, 64.2cm by 111.8cm, George Fennel Robson.

Figure 2.2 *Fishing by Torchlight*, 1817, 17.5cm by 28cm, Joseph Lycett.

Figure 2.3 *Yosemite Valley from the Best General View*, 1866, Carleton Watkins.

2.1 Valuing landscapes

How can we best describe the sort of value that we are thinking about here? If we are interested in the look of a thing, and not in what it can do for us, we are interested in its *aesthetic* properties. Talk of **aesthetics** is often connected with art, but is relevant to much else besides, including the other things we make (buildings, motorbikes, coffee machines), the things we don't make (icebergs, tigers, skies) and of course human beings themselves (Greta Garbo, Brad Pitt). These examples all focus on the way things look, but we might be concerned as well with sound (music, birdsong), smell and taste (newly mown hay, food) or with feel (stroking a cat, crunching through snow). And, though the examples so far may seem to be linked with beauty, a concern with aesthetics and aesthetic value can just as well focus on ugliness, dullness or convolutedness, and much else besides. There are some beautiful landscapes, but there are more that are ugly, messy, boring or garish. All of these are aesthetic judgements as to the landscape's worth.

aesthetics

If we are considering something from the point of view of its aesthetic properties and trying to assess its aesthetic value, we are interested in it not as a resource, or means, but as an end. We are interested, then, not in its instrumental but in its non-instrumental value. And while many aesthetic values are subjective or personal (some people like the smell of the farmyard or the slimy feel of fresh trout, while others hate them both), others appear to be objective, and so relating to things that we ought to value. So it is with many landscapes. Even if you prefer life in the city, you'll probably agree that the kinds of landscapes shown in Figures 2.1 to 2.3 are high on most people's lists of beautiful places, places that many think we need to protect. Governments seem to believe this too when they designate certain places as Areas of Outstanding Natural Beauty (AONBs), thereby seeking to protect them from development and the wrong sort of change. Similarly, many think that ugly, neglected or abandoned landscapes are things that we might, with good reason, want to improve. Again, governments, in various of their policies, seem to agree. Environments matter, and for many reasons; but that they are beautiful is important among these.

For more on the distinctions between instrumental and non-instrumental value, and subjective and objective value, see **Hinchliffe and Belshaw, 2003**.

The view that beautiful landscapes are genuinely valuable, things for which many of us are and ought to be concerned, can seem straightforwardly true to some, while deeply problematic to others. Discussions of value inevitably become complex. And as a first step to furthering our understanding here we can consider a number of questions about such beauty, along with some answers that are often given:

- First, is it really obvious that there's beauty in such landscapes? Some people think it's extremely obvious, that beauty is staring us in the face. The philosopher Terry Diffey, for example, has commented that 'we can all think readily enough of examples of natural beauty: a prospect, say, of the Sussex Downs as viewed along its line from the top of one of its hills is *undoubtedly* beautiful' (Kemal and Gaskell, 1993, p.48).
- Second, how widespread is the appreciation of landscape? Some have insisted it is very widespread. The art historian Kenneth Clark (1949) has claimed that

'with the exception of love, there is perhaps nothing else by which people of all kinds are more united than by their pleasure in a good view'.

- Third, how is this near-universal appreciation of particular forms of natural beauty to be explained? It seems that very many of us are attracted by roughly the same sorts of things. Is that just a coincidence, or can we account for this apparent conformity in our tastes and preferences? A recent and influential suggestion (Appleton, 1996; Muir, 1996) has been that our appreciation of landscape derives from basic human drives, and that the sorts of creatures we are – our emotions, our needs, our history – all help determine the sorts of things we like. So, the story goes, the likely origin of our species in the savannahs of East Africa plays some kind of explanatory role in many of our preferences. We appreciate wide expanses of grassland, scattered trees, the play of light on water, because our ancestors began and developed in such places (Figure 2.4). This *habitat theory*, then, connects our shared aesthetic response to landscape with our shared evolutionary background.

Figure 2.4 The African savannah. Our ancestors evolved in environments such as this and there are those who argue that the almost ubiquitous human appreciation of environments is born out of our ancestral evolution.

Activity 2.1

Think about these claims. Various objections can be made. One is that people often disagree about whether or not landscapes are beautiful. Another is that many beautiful landscapes are destroyed to build new towns or motorways or business parks. A third is that many favourite landscapes don't look much like savannah. But are these objections powerful? Consider what you might say in favour of the claims and against these objections.

Comment

These objections are not, I think, as strong as they may seem. Certainly there's conflict and disagreement about beauty and about values more generally. But the mere fact that people disagree doesn't itself show that this beauty, these values, don't really exist. People disagree about God and the Loch Ness Monster. But that doesn't show they don't exist. They disagree too about nuclear weapons and how murderers should be punished, but this doesn't overturn the view that some weapons and some punishments are inappropriate for a civilized society. When Diffey says the Downs are undoubtedly beautiful, he might just mean that they *really are* beautiful, even if some people might not see it.

Nor does the fact that we often destroy beautiful landscapes upset anything I've suggested. We might think the view down a valley is genuinely beautiful, get pleasure from looking at it, and yet still decide to flood it so we can supply a city with water. We may decide the instrumental, resource value of water outweighs the non-instrumental aesthetic value of the landscape in its present state. Different values can conflict even though both are real.

Nor, finally, does the fact that only relatively few admired landscapes look much like the African savannah demand that we throw out habitat theory lock stock and barrel. We can't expect total conformity. And a closely related account, *prospect-refuge theory* – suggesting that there's some match between the landscapes we like and those that offer scope for a hunter-gatherer to capture food and to see without being seen – has more flexibility and more hope of success.

Yet even if the criticisms put forward in the above activity don't immediately overturn the suggestion that landscapes can be objectively valuable (rather than just something for which some do, while others don't have a taste), it may be that if we delve deeper we'll find reason to doubt the objectivity view.

2.2 The culture of nature

Even if there is such a thing as natural beauty, there's perhaps not much that is natural in our appreciation of it. Take one example: the discovery of the English Lake District as a place to visit and marvel at, a place in which to admire the beauties of nature. This discovery occurs not long before William Wordsworth moved into Dove Cottage, between Grasmere and Rydal, in 1799 (see Figure 2.5). It was only towards the last quarter of the eighteenth century that this corner of northern England began to be seen as offering a convenient, and much cheaper, alternative to that fashionable eighteenth-century phenomenon, the European Grand Tour. Although the lakes, mountains, woodland, had long been known about, they had in general been viewed with suspicion and hostility. And this change in attitudes was remarkably sudden. Daniel Defoe, writing in the 1720s, describes the region as a 'barren and frightful' place, to be got through as speedily as possible, and in 1807 the poet Southey wrote, though tongue-in-cheek, of 'tracts of horrible barrenness, of terrific precipe, rocks rioting upon rocks, and mountains tossed together in chaotic confusion'. So perhaps we shouldn't speak of the discovery of beauty, so much as a change in taste, or the invention, within the newly formed Romantic imagination, of quite different notions as to what natural beauty is (see the extract from *The Prelude* by Wordsworth, in Section 2.3).

Something akin to this is at work as well in the USA, when, around the middle of the nineteenth century, and influenced in part by the writings of Henry David Thoreau and Ralph Waldo Emerson, the discovery of the American landscape (particularly the mountains, first of New England and later the Rockies and the Sierras) seemed to promise not merely material but also spiritual benefits to those brave enough to leave the cities and farms behind and venture into the unknown. Artists and photographers like Thomas Cole (see Figure 2.6), Frederick Church, Albert Bierstadt and Carleton Watkins did their best to capture, and even improve on, the immensity of these mountain landscapes, bringing their works, and reproductions of their works, to a vast and less adventurous audience in the urban and suburban centres of the east and south.

Similarly too, and around the same time, in Scotland, Wordsworth's near contemporary, the novelist Walter Scott, played a significant part in cultivating a taste for the Highlands, and Queen Victoria's well publicized travels, along with Landseer's and others' paintings (see Figure 2.7), all encouraged this further. Those moody, misty autumnal views, highland cattle by the side of a loch, distant glimpses, in the evening sky, of further mountains or the view out to sea and the islands, are still familiar imagery. They may be viewed, not as an instinctive and

Figure 2.5 Grasmere. The Lake District became a fashionable retreat only towards the end of the eighteenth century.

Figure 2.6 *Landscape with Tree Trunks*, 1828, 67.3cm by 82.5cm, Thomas Cole. Towards the end of the eighteenth century, paintings such as this were bringing the pleasures of the American landscape to the American public.

Figure 2.7 *The Windings of the Forth*, 1758–1840, 73.7cm by 45.7cm, by Alexander Nasmyth. An example of the type of painting that did much to popularize the Scottish Highlands.

untutored response to the beauties of nature already extant, but as the product of cultural change and as offering a much needed refuge or escape from what many saw as an increasingly industrialized, mechanized and in the end dehumanized urban existence.

Activity 2.2

Look again at Figures 2.1 to 2.7. How might someone argue that landscapes like these give evidence not so much of natural beauty as of one culture's construction of the image of beauty?

Comment

There are a couple of things to observe here, some of which you might have overlooked because they are so obvious. First, you're looking in all these cases at pictures of nature, rather than nature itself. It's nature interpreted, even in the case of photographs, by artists. Second, the names and dates suggest that these images are not taken at random from around the world and at different times, but are likely to be British or North American works of the past couple of centuries. The artists knew, and were influenced by, each other, and were responding in their own work as much to further images as to nature itself. Third, looking at and thinking about landscapes like these might then remind you that the term itself is ambiguous, referring both to works within a genre of painting and to parts of the surface of the Earth. It isn't, though, merely a coincidence that the term has these two meanings, for painting landscapes develops very much around the time, and in the place, where an awareness of and aesthetic response to the natural landscape first becomes widespread.

A fondness for certain sorts of landscape is instinctive to many of us. It can be tempting to follow the suggestions aired above and think that such appreciation is wired in. Yet as we've now seen, rather than occurring spontaneously, as a part of human nature, this taste is in many ways invented, written up and handed down. And it's impossible to know for certain how stable it is or how long it might last. There's already something seemingly old fashioned about much of this nineteenth-century fashion for wandering about in the mountains. It may be that as with clothes, cars and home furnishings, tastes in landscape will change.

2.3 Varying the scene

Yet there are landscapes and landscapes. And it is a good idea to make some distinctions between them, even if we might think of all of them as beautiful and in different ways as worth producing or protecting, and even if one kind of landscape merges into another.

First, there's the kind we have focused on so far, the dramatic and wild, the overbearing mountains and brooding immensity of the darkening sky. These are the landscapes that figure in the paintings of Turner and some parts of

Wordsworth; landscapes the taste for which is especially associated with **romanticism**. Is there really beauty here? We may say so, but particularly at the close of the eighteenth century, when this taste began to develop, people referred more often to the **sublime**, and wanted to contrast these overwhelming and sometimes even frightening scenes (Figure 2.8) with the more familiar and more obviously beautiful. Of course, such landscapes are not often genuinely frightening, but they offer the sort of pretend fear we might, today, get from horror films or fairground rides, and so provide a kind of escapist entertainment for the jaded city dweller.

romanticism

sublime

Extract from *The Prelude*

A rocky Steep uprose

Above the Cavern of the Willow tree

And now, as suited one who proudly rowed

With his best skill, I fixed a steady view

Upon the top of that same craggy ridge,

The bound of the horizon, for behind

Was nothing but the stars and the grey sky.

She was an elfin Pinnace; lustily

I dipped my oars into the silent Lake,

And, as I rose upon the stroke, my Boat

Went heaving through the water, like a Swan;

When from behind that craggy Steep, till then

The bound of the horizon, a huge Cliff,

As with voluntary power instinct,

Upreared its head. I struck and struck again,

And, growing still in stature, the huge Cliff

Rose up between me and the stars, and still

With measured motion, like a living thing,

Strode after me ...

(William Wordsworth, The Prelude, 394–412; in Gill, 1984)

A second sort of escapism has a longer history. The notion of the sublime developed in the later eighteenth century, in part to complement and contrast with another kind of landscape that was, in earlier times, more carefully and accurately described. For there was, especially in northern Europe, a widespread taste for the **picturesque**, the sort of landscape that features meandering streams, a ruined cottage, often a gnarled tree in the foreground and perhaps a shepherd or a milkmaid, or both, disappearing into a distant copse. This is the sort of landscape that might be associated with the Dutch painter Ruysdael, or the Italian, much admired in England, Salvator Rosa, or even perhaps Constable (Figure 2.9) The taste here goes back to the seventeenth century and reflects the bourgeois values that were then coming to predominate, especially in Protestant countries like England and Holland.

picturesque

Figure 2.8 *An Avalanche in the Alps*, 1803, 109.9cm by 160cm, Phillippe Jacques de Loutherberg. The sublime landscape promised 'a sort of delightful horror, a sort of tranquillity tinged with terror' (Edmund Burke, quoted in Andrews, 1999, p.134).

Figure 2.9 *The Hay-Wain*, 1821, 130.2cm by 185.4cm, John Constable. Familiar now but revolutionary then.

A third kind of landscape extends further back still. Often the concern here is with a more obviously tamed and ordered nature, in which trees, crops and fields are well groomed and cared for, and viewed in a clear and detail-revealing light. Often too, brand new buildings, and not just their ruins, will figure within the scene. And, in contrast to the picturesque landscape (which might include a handful of human figures, usually rustic types, as just one element in the composition), these earlier views will give people much more prominence, with the natural landscape itself playing a background role. And these people, who often form the real subject of the painting, are less likely to be peasants or farmhands than the owners of the land behind them. You might associate this more **pastoral**   pastoral landscape in the eighteenth century with Gainsborough (see Figure 2.10), but it can be readily discovered in painting from the fourteenth century on (see Figure 2.11). A consciousness of the pleasures in such landscape can be traced further back still – the Italian poet Petrarch was the first person known to climb a mountain simply for the view, while delight in the settled and settling nature of life in the countryside is evident in writings from ancient Greece and Rome (see Figure 2.12). There is a geographical as well as a historical dimension to consider here. Again we can learn from looking at art. In India, the Islamic world and in much of east and south-east Asia, sculpture, pottery and painting often reveal a delight in the pastoral landscape, the image of nature subdued, ordered and brought under human control for human benefit (see Figure 2.13). And it is surely worth noting (as you'll see in comparing Figure 2.10 with Figure 2.4) that not only is it the most widely appreciated, but it's landscape of this kind that fits best, though still far from perfectly, with the suggestions of habitat theory.

Figure 2.10 *Mr and Mrs Joseph Andrews*, 1750, 69.8cm by 119.4cm, Thomas Gainsborough.

Figure 2.11 *Allegory of Good Government: Effects of Good Government in the Country*, a fresco from the west wall of the Sala della Pace, Palazzo Pubblico in Siena, 1338–40, Ambrogio Lorenzetti.

Figure 2.12 A fresco from the wall of a house in Pompeii – the Casa del Bracciale d'Oro, AD79.

Figure 2.13 Moghul Indian palace garden court, c. seventeenth century, 33cm by 21.5cm.

2.4 Beauty

Where does this leave the environment? We began with a straightforward position, that there is beauty in nature that most of us recognize and admire. This, it can be argued, gives us reason to preserve at least those parts of the environment that are beautiful. It gives us reason also to alter those parts of the environment that aren't beautiful, the parking lots, urban and suburban sprawl and industrialized farmland.

But now this position has been variously complicated, for what has been suggested is three things. First, that many of our ideas about natural beauty need to be learned, and often reflect our social position (either as an owner of land or as a city dweller with fantasies of the country life). Second, because of this, there is often conflict and disagreement about what is beautiful and what is not. Finally, attempts to tie our notions of beauty with our shared ancestry are less straight-forward than they first might seem. But then what happens to the idea that some landscapes might be really, objectively beautiful and so objectively valuable, and so, in turn, things we ought to care about and seek to preserve for future generations? Doesn't it look as if the old saying, 'beauty is in the eye of the beholder', is right after all and that our attempts to argue about and defend landscapes are just, in the end, efforts to impose our tastes and preferences on others?

A number of points can usefully be made here. First, it's worth resisting a simple answer to the question of where beauty is located, whether it's in the mind, or in things. Instead we should think that beauty, like colour, depends on the relation between the two. I can say grass is really, objectively green, even though it doesn't appear green to moths, and might not appear green to Martians. Let's just say it's human-green. In a similar way Greta Garbo was human-beautiful. Someone who thinks grass is red, or Garbo ugly, has just made a mistake. Second, and following from this, we might think, just as with colour discrimination, that some sort of aesthetic appreciation of nature is normal, natural and even beneficial for human beings. Because we have reason to believe that future generations of humans will appreciate nature in some form or other, so we have reason to ensure that fair-sized chunks survive. And we have good grounds for arguing with those who think that, so far as its aesthetic value is concerned, this doesn't matter at all. Third, we might argue as well that we do have reasons to preserve particular forms of nature, and particular kinds of landscape. It might not be straightfor-wardly natural to like mountains, it might not in any way be built into our genes, but if such liking emerges in a certain culture and we have reason to think similarly shaped cultures will exist in the future, then again we have reason to think a liking for mountains will persist. I think it's reasonable to believe that many people in the future will live in cities and will need, from time to time, to get away from it all. If I'm right about this, then much of the objectivity view remains: there is beauty in nature and there are reasons to look after the kind of scenery we have today, even if, for a short time at least, it seems to us to be unimportant.

Summary

Though almost everyone agrees that there are some beautiful landscapes, there is often disagreement both about which landscapes are beautiful and about the kind of value that such beauty may have. This disagreement matters. Conflict over what is beautiful leads to dispute about how to rank different kinds of landscape. Conflict over whether beauty is of any genuine value at all threatens landscape preservation whenever other values enter the fray.

The simple view that landscapes such as those in Figures 2.1 to 2.3 are straightforwardly and uncontroversially beautiful has to be rejected. We need to accept that many of our ideas of what is beautiful in nature need to be learned, are subject to cultural bias, and often reflect our social position either as an owner of land or as a city dweller with unrealistic notions of the country life. Disagreement about landscape values is inevitable, and has no hope of always being resolved.

Nevertheless, there is reason to believe that, in broad terms at least, an appreciation of landscape, indeed natural beauty in general, is a desirable and near-universal trait. And so aesthetic values, even taken alone, give us reason for preserving at least some landscapes for future generations.

3   Parks

To value landscape is to value the look of nature, or to value painting. There is, as I've already suggested, an ambiguity here. But in spite of the differences, important similarities remain. For, whether you're looking at paintings at home, in a gallery or shop, or at the real thing from a hotel balcony, a pleasure boat or through the window of a car, coach or train, to value landscape is to stand outside of nature, and then to select and frame certain aspects within it, constructing divisions where, in reality, none exist. The boundaries to paintings and photographs are real enough, and linked closely at the height of the landscape fashion with a series of devices to limit and direct the gaze. So the eighteenth-century tastes for the picturesque and the sublime went hand in hand with the use of the camera obscura, the Claude glass (see Figure 2.14), and the so-called Gilpin tints by artists and tourists alike, in order to collect, contain and colour the landscape, to shrink it to a more manageable size, deepen the blues or greens or make more dramatic the sunset. A modern equivalent here is the digital camera where, even before a picture is taken the landscape can be gathered and flattened on to a tiny screen, altered in various ways, then made into something you can take away, own and enjoy again at a later time.

But there is more to an aesthetic appreciation of nature than this. Increasingly, as the nineteenth century went on, many people grew to be dissatisfied with merely looking at nature. They wanted to be within it, to move about in it and smell, listen and touch it. Hostility to mere scenery, to nature at a distance, is often evident in the work of writers such as William Wordsworth and John Ruskin, and, in America, Emerson and Thoreau. There's a complex mix of motives in play here, with many in the nineteenth century still believing that the more closely the beauties of nature are penetrated, the more they will reveal the hand of God within them. This overt religiosity, along with any widespread tendency to scientific curiosity, may by now have dwindled, but still many people will claim to be uplifted, feel themselves more whole, if they can linger and wander within the countryside. And it is in large part from the combination of this wider appreciation of nature, and an awareness of the threats posed to it by the increasingly voracious appetites of industry, business and the spread of cities, that the idea of national parks began to take shape.

Figure 2.14 A collection of Claude glasses and tints, used in the eighteenth century for creating photographs.

3.1 National parks: origins and backgrounds

Perhaps mention of parks still conjures up, more than anything, the town or city plot (Central Park in New York, Edinburgh's Waverley Gardens, the Bois du Boulogne) often developed in the middle or late nineteenth century when civic pride and the wealth and power of the corporation were at their height, in order to provide some respite from an unrelenting urban existence. But this sort of park, relatively small (Central Park covers roughly one square mile), its boundaries evident, its emphasis clearly on recreation – the bandstand, formal gardens, tennis courts and putting greens – is, in the longer view, very much the exception, and for most of their history parks have had to occupy a more complex and diverse set of roles and functions. So as well fulfilling some aesthetic purpose, parkland has been seen as a resource for hunting, forestry and agriculture, and has often been designed to reflect and enhance the status of its typically aristocratic owners. Boundaries are often ill-defined, as park merges into farm or the edges of towns and villages beyond. And only rarely has it been cleared completely of human inhabitants, even though their values, and those of the owners, were often in conflict.

National parks, in many parts of the world, have more in common with the country estate than with recreation areas in the city. Covering vast areas, their perimeters often unmarked, with human beings and human artifice taking at least a secondary role, the emphasis is on nature and nature's preservation. Yet parallels with the city park cannot be denied. Recreation, though often of a subtler kind, has long been high on the national parks agenda. While city parks reflect local prosperity, their national counterparts are often intended to foster pride in the nation, nowhere more so than in the USA, where the perceived absence of a self-standing human culture led to the promotion of nature in its place. The redwood forests, the mountains framing the Yosemite Valley (see Figure 2.15), Niagara Falls, and other wonders besides, were all frequently compared with European cathedrals, and travel to such places was often seen as the New World equivalent of the Grand Tour.

City parks are neither developed nor sustained without argument. Yet in many cases, and certainly among the most famous, there is broad agreement as to their existence and use, and it's hard to see how their future, in much the same shape as their present, can be seriously in doubt. The situation with national parks is considerably less secure. Given their size and the alternative possible uses of land (agriculture, intensive forestry, mineral extraction), given too their often ambiguous relations with the locale, and, finally, given their emphasis on nature conservation, which often seem at odds with the promotion of human values, it is hardly surprising that there are widespread and longstanding disagreements about their purpose, their management, and even whether they should exist at all (Figures 2.16, 2.17 and 2.18). Even if substantial numbers do now support the view of the national park as some sort of nature reserve in which human needs take a back seat, the consensus is fragile and the route to consensus, over the past hundred years or so, more circuitous than much of the promotional literature

Figure 2.15 A nineteenth-century stereograph of Yosemite Valley: a mass-produced image of splendour for the tourist market.

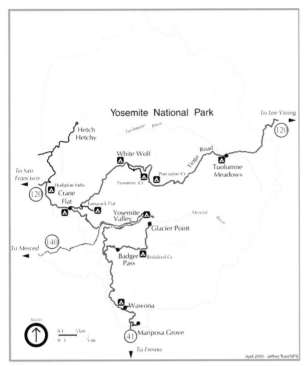

Figure 2.16 Map of Yosemite National Park, USA, approximately 3,975 square kilometres.

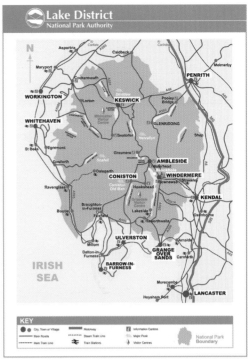

Figure 2.17 Map of Lake District National Park, UK, approximately 2,292 square kilometres.

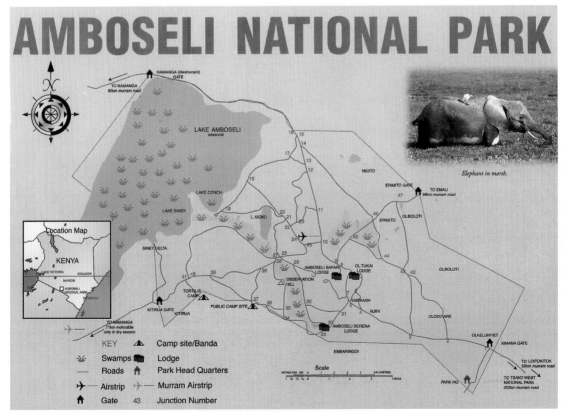

Figure 2.18 Map of Amboseli National Park, Kenya, approximately 392 square kilometres.

would have us believe. Yet generalization here can be suspect, and the world's national parks still vary enormously in size, design, history and function. We can't consider them all, but the tables below give you some outline information about a sample of three.

Table 2.1   Lake District National Park

size and location	Cumbria, in north-west England, approximately 866 square miles/2,292 square kilometres.
human population	c.42,000.
non-human population	Noted for its sheep, bracken, mixed woodland.
origins	Mooted by Wordsworth, in his 1835 guide, when he saw the lakes as 'a sort of national property, in which every man has a right and interest who has an eye to perceive and a heart to enjoy'. Impetus sustained with land acquisition by the National Trust. In 1935 a Standing Committee on national parks lead to the 1949 National Parks and Access to the Countryside Act, which lead, in turn, to the park's formation in 1951.
famous names	Wordsworth, John Ruskin, Beatrix Potter.
ownership and management	Almost all land in private ownership (including National Trust). Managed by Lake District National Park Authority, in conjunction with local and central government bodies.
slogan/purpose	(a) To conserve and enhance the natural beauty, wildlife and cultural heritage of such areas. b) 'To promote opportunities for the understanding and enjoyment of the special qualities of these areas by the public' (1995 Environment Act).
major challenges	Changes in farming practice.

Table 2.2   Yosemite National Park

size and location	Northern California, in the USA, approximately 1,500 square miles/3,975 square kilometres.
human population	c.1,400.
non-human population	Bears, deer, redwoods.
origins	Yosemite Valley a State Park in 1864. Yellowstone the world's first national park in 1872, with Yosemite following in 1890. The valley incorporated into the park in 1906.
famous names	John Muir, Teddy Roosevelt.
ownership and management	Land owned by the Federal Government.
slogan/purpose	To conserve the scenery and the natural and historic objects and the wild life therein and to provide for the enjoyment of the same in such manner and by such means as will leave them unimpaired for the enjoyment of future generations.
major challenges	Tourist numbers – conflicts between recreation and conservation/preservation.

Table 2.3 Amboseli National Park

size and location	At the foot of Mount Kilimanjaro, in south-east Kenya, approximately 185 square miles/392 square kilometres.
human population	Officially, none, but (a) some incursions by pastoralist nomads are tolerated and (b) populations just outside the park have a serious impact upon it.
non-human population	Elephants, ungulates, occasional lions.
origins	Over-hunting led to game reserves, first for regulated hunting, later for preservation and tourist purposes; the Park was created in 1973.
famous names	Richard Leakey, Daniel Arap Moi.
ownership and management	Kenya Wildlife Service
slogan/purpose	To work with others to sustainably conserve, protect and manage Kenya's invaluable biodiversity for the benefit of Kenya and as a world heritage.
major challenges	Under-regulated tourism, agriculture, land-use conflict.

Activity 2.3

From the information given here, along with any background knowledge you might have, what strike you as the major differences between these parks?

Comment

There can be no single right answer to a question like this. However, considering the parks in pairs, the main differences seem to me to be these:

- *Lake District/Yosemite*: the former isn't owned by the government, and it has a big human population. The latter is much more of a state-owned and managed wilderness reserve.
- *Lake District/Amboseli*: relations between the park and human populations within or near to it are seemingly less fraught in the former than in the latter.
- *Yosemite/Amboseli*: it appears that while both are trying to be managed wilderness areas, the former, for some reason, can look forward to a more secure future than can the latter. One relevant difference is that migrating animals are much more important to Amboseli than to Yosemite. Notice also the vast difference in size.

3.2 Contesting the park

There are already issues surfacing here, about the organization of parks, and the management of potential conflict situations that arise within them. We can consider some of these issues in more detail, by looking at the different sorts of threats or challenges to a park's organization.

Human beings

We can focus on problems of two kinds here. First, there are the demands and needs of the park's original inhabitants, many of whom have either resisted the setting up of the park in the first place or who have since come to believe that its interests and theirs are fundamentally at odds. This kind of conflict is most evident today in Africa. Its park system reflects European values (first hunting for sport and, later, conservation and tourism) and has mostly been managed by white Europeans. The native African voice has been too little heeded. Both in the Amboseli park and also on the virtually adjacent Maasai Mara Game reserve there's been a long history of misunderstanding and antagonism, especially between the park authorities and the Maasai, pastoral-nomadic people who have traditionally made use of lands from which they are now excluded (see Box 2.1). Conflict here has frequently resulted in violence, with the Kenya Wildlife Service under Richard Leakey initiating a shoot-to-kill policy against poachers (see Figure 2.19) and with the Maasai killing lions and destroying a water pipeline in exasperation at their increasingly marginalized position.

Such conflict might at first seem a non-issue in a place like Yosemite, which would appear to be fundamentally a wilderness area. But this is only because the USA government, in a not untypical example of ethnic cleansing, was so successful in removing its original inhabitants, the Ahwahneechees, to reservations beyond the park's boundaries. The Lake District is different here. Its local inhabitants could hardly be shipped off elsewhere, or obliged to eke out an ersatz rural existence simply in order to satisfy the tourist gaze. They are English-speaking English citizens, in an area with a long-standing mixed economy (forestry, agriculture, quarrying, mining, light industry) that has been interwoven

Figure 2.19 Kenyan Wildlife Service officer destroys illegal ivory captured from poachers.

Box 2.1   Interviews with Maasai farmers

Has the creation of the Amboseli National Park been a good thing or a bad thing for the Maasai?

Olurei: It's a very bad mistake. The park is where all the animals are in the dry season because of the swamps and the river. But then in the rains all the animals are out and eat our grass, and only go back into the park in the dry season. The wild animals have a grazing reserve but we don't ... When one of my friends tried to graze in the park he was grabbed, jailed and fined heavily. They forget the park was made out of our land.

Do you agree with conservationists who say that wildlife and livestock compete in the Mara?

Ntoros ole Baari: The land could easily accommodate both the livestock and wildlife. There are as many wild animals on the Maasai lands as in the reserve. Livestock and wildlife are complementary. ...

What has happened to the Maasai kept out of the Mara?

Ntoros ole Baari: The Maasai are forced to overgraze as so much of their land has been reduced by conservation. By keeping the Maasai out of the reserve, conservationists are in fact contributing to the threat to the environment.

The authorities have argued that the Maasai have had to leave Amboseli because they were causing too much damage ...

Metoe ole Loombaa: No. The Maasai have coexisted with wildlife for a long time and have never caused any damage. We don't eat wildlife, so we never hunt. The manure our cattle were leaving in the park was making the vegetation grow. Now it's completely destroyed and the animals are following us to our ranch to follow our livestock. The park was better off when we were there – you could find rhinos and lions then. But now all the animals have moved out.

(from an interview by Ruth McCoy; quoted in Trillo, 1999, pp.376–7)

seamlessly with that of the country as a whole. This is the major difference between the English park and those in the USA and Africa. The Lake District was never intended to be a large-scale nature reserve or wilderness area. This isn't to say conflict is unknown, and the difficult terrain of the Lake District, combined with demands to preserve its scenic aspects, has led to significant and unwelcome delays in providing services, most notably electricity, to inhabitants in the remoter valleys (Berry and Beard, 1980).

The second kind of problem involves the conflicts between the various demands that outsiders can bring. In the Lake District there are disputes among different tourist interests, most notably at Windermere, where some groups valuing peace and tranquillity want to stop power-boat activity on the lake. Such disagreement about recreation is today less evident in Yosemite (see Figure 2.20), but it has been an issue in the past. It was only in the 1950s that the nightly tourist attraction of heaving burning coals, woods and fireworks over Glacier Point came to an end, and as recently as the 1970s that plans were unveiled for a new hotel in the same place, linked by tramway to the valley floor. Developments such as these may seem unthinkable in the present conservationist climate, but climates do change. Similar problems in Africa spill onto the world stage, with the need to find a balance between the global tourist's demands for access to animals, and the perceived need by international conservationist groups to minimize disturbances. In all such cases money talks: though ballooning will frighten the zebras, at $250 per person for a 45 minute trip, tourist demands are not easily resisted (see Whelan, 1991).

Figure 2.20 Recreation in an American national park. *The Trip West. Camp Ground No.4, Yosemite National Park,* 1966, Bruce Davidson.

Boundaries

All environments impinge upon both their immediate neighbours and, although often to a lesser extent, on those further away as well. National parks cannot be effectively ring-fenced, and both the legal and physical barriers created to protect them will remain porous. Certain of these external threats derive from identifiable sources, while others are harder to pin down. The Lake District was noticeably affected by the Chernobyl disaster of 1986, with both meat and milk showing marked contamination, and there remain well-founded fears both about the risk of serious accidents in the future and about the obfuscation of ongoing problems at the nuclear reprocessing plant at Sellafield, only a handful of miles from the park's boundary (see **Hinchliffe and Blowers, 2003**). Many of the parks in Africa, Amboseli among them, are highly susceptible to the effects of climate change, with a cycle of floods and droughts, exacerbated in recent years by the El Niño effect, unlikely, for the foreseeable future, to recede. Only concerted international action, it seems, can do anything of significance to counter global warming, and such action remains a long way off. The question for park authorities, in such circumstances, is whether they should attempt some kind of holding operation – and this is often what the public wants – or whether they should allow nature to take its course. Amboseli is threatened in a further way (and one which distinguishes it from Yosemite, where most of the attractions are stationary), for an estimated 80 per cent of its animal populations migrate regularly beyond the park's boundaries. Without sufficient attention being paid to the entire Amboseli ecosystem – only 6.5 per cent of which lies within the park itself – wildlife populations can be unsustainable.

Resources

Water is often a contested resource in parklands (see Chapter Four). Our affection for mountains ensures that many of the world's parks are situated in areas of high rainfall. The need to protect watersheds often plays a significant motivating role in the park's creation. But then this same topography encourages the development of reservoirs, either from scratch or by enlarging pre-existing lakes, typically at some perceived cost to the park. Thirlmere, in the Lake District, was dammed in 1890 as a result of increasing pressure from the industrial and fast-growing Manchester conurbation some 50 miles to the south. And while Yosemite valley itself remains in this respect untouched, only a few miles away the park's second major valley saw the creation of the Hetch Hetchy reservoir in 1913 (see Figure 2.21). This was built to serve the needs of San Francisco which, after the devastation of the 1906 earthquake, launched enthusiastic rebuilding and expansion projects. It has been argued that the preservationists' earlier successes helped to seal Hetch Hetchy's fate: had this valley, like its neighbour, been made more accessible to the ordinary tourist, then public pressure in favour of its protection would probably have been more pronounced.

> Dam Hetch Hetchy! As well dam for water-tanks the people's cathedrals and churches, for no holier temple has ever been consecrated by the heart of man.
>
> (John Muir, quoted in Opie, 1998, p.389)

(a)   (b)

Figure 2.21 Hetch Hetchy Reservoir (a) looking up the valley before the creation of the Hetch Hetchy reservoir in 1913 and (b) the same view today.

Are these 'threats' really threats?

Values are implicated in all talk of threats. Whether an anticipated change actually threatens a thing depends on whether, on balance, we judge that change to be for better or for worse. Many people demonstrate, perhaps with good reason, a strong tendency towards conservatism where national parks are concerned. Yet it is worth asking whether this hostility to change is always well founded. I'll mention just two examples.

First, we might think further about whether the damming of Thirlmere was, so far as the Lake District is concerned, a bad thing. It was a good thing for Manchester, of course, and that may be reason to think it was a good thing overall – a utilitarian or cost–benefit approach (see Chapter Seven) might demonstrate that losses are clearly outweighed by gains – but conventional wisdom has it that lakes are better than reservoirs. Yet a comparison of Thirlmere with Windermere just to the south (and there were proposals to dam Windermere as well) shows that this cannot be straightforwardly true. Although when levels are low a reservoir can be a barren and depressing place, the need for purity of supply has ensured that the area around Thirlmere is free of virtually all development. And the planting of a mix of trees, most of them non-native conifers, dense but increasingly unregimented, gives the area a Gothic sublimity that is nowhere rivalled in England (see Figure 2.22). This is an altered nature, perhaps, but it isn't obviously any the less moving for that.

Figure 2.22 Thirlmere: fake nature?

Second, the claims made by the Maasai people to make use of parklands from which they've been excluded need not threaten the park or, more particularly, the sustainability of its wildlife populations. Given human population increases, controls would be needed, and there is no recoverable golden age when human and non-human species lived in unregulated harmony. But while many tourists seem to prefer a land wholly free of people, not only is there no historical warrant for this, but there is often reason to believe that the park's ecosystems may actually benefit from moderate human involvement.

3.3 The tame and the wild

Just as our responses to nature turn out to be less natural than they first appeared, so too both landscapes and parks often deceive us as to how natural they are. With pasture land, forestry, the well-placed copse, mountains denuded of their trees by sheep, almost every square mile of the Lake District is shaped by human endeavours. Some of these are recent, as with Thirlmere, some of them ancient, as in the coppiced woods, dating from the Middle Ages, behind Ruskin's house near Coniston. Even where parts of the landscape do look something like they did before human habitation, this is often the result of strenuous human involvement: replacing non-native trees with native species (see Box 2.2); removing failed sea defences from coastlines; and introducing complex wildlife management schemes.

> ### Box 2.2 Native and non-native trees of Britain
>
> *Some native trees*: willow, birch, scots pine, bird cherry, hornbeam, box, beech, oak, elm, ash, wild service tree, whitebeam, field maple, yew, hawthorn, holly, crack willow, aspen, bay willow, rowan, sallow, juniper, hazel.
>
> *Some non-native trees*: sycamore, cedar, rhododendron, larch, spruce, walnut, fir, chestnut, plane.

Another question about beauty needs to be looked at here. I said earlier that if we are interested in a thing's aesthetic properties, then we are interested in the way it looks, in its appearance and not in its use or its history. But is this really right? Is beauty, as this suggests and is commonly believed, really only skin deep? Or is there more to it?

Think first about people. Though increasing numbers of people spend time and money in altering their appearance, the reaction of many is still to look upon the product of artifice as unreal, to think that someone is not really beautiful if their seeming beauty is the result of drugs or corsets or knives. We seem to be concerned not only about how people look, but also about how that look came about.

The same is true of our responses to the environment. What many people believe is that we ought to be concerned with more than the mere appearance of things. So the sweet smell of spring is less sweet, many believe, if it turns out to come not from the hedgerows but from a deodorant factory just over the hill. A beautiful flower may be thought less beautiful if its shape or colour is the result of interference or disease. Where landscape is concerned, similar considerations come into play. We believe the rocks of the Lake District are among the oldest in Britain, its highest mountains produced millions of years ago, its valleys gouged out by the unyielding flow of ice. We think that Amboseli, like most of the other parks in Africa, wears a similarly ancient face, and that both the people and the animals upon it are living as their ancestors did, in an uninterrupted chain. But what if these things turn out not to be true? Imagine we were to discover that seemingly age-old hills have been recently shaped up by bulldozers, and covered with imported grass and turf? Or we find there are no longer any genuine hill farmers and shepherds, living the simple life as their ancestors have done, but only newcomers who dabble in farming with EU grants, their 'cottages' looked after with the help of the National Trust, their quad bikes and satellite dishes hidden from the tourist view? Or again, that even in seemingly wild places the 'balance of nature' is artificially sustained, with culling and breeding programmes to preserve what is thought of as equilibrium? We discover these things and we would value the landscape and the park differently. And perhaps we would value it less, thinking it less beautiful, less awe inspiring, less sublime.

But are we right to adjust our evaluations in these ways? Does fakery, as at Thirlmere, really matter? There are different views, and different arguments here. On the one hand, some people believe that whether or not a thing is beautiful depends solely on the way it looks, while others insist that detailed understanding of the science involved – knowledge about geological time, biological processes, the extent to which an environment is natural or human-made – is essential to a proper aesthetic evaluation. I'm taking no sides here, but simply pointing out another area where talk of 'the beauties of nature' is less straightforward than at first it may seem.

3.4 Conflicting values, subterfuge and spiritual needs

No park can be without contest. But contest exists to different degrees, and comes in different kinds. In the Lake District, even though different groups have their different visions of the park's future, and will weight its assets in different ways, there is general agreement as to what those assets are, and agreement too that the park should continue into the future in something reasonably close to its present form. Similarly for parks in the USA, but with the important difference that groups who, because of their different cultural background, might have had a radically different vision of our proper relation with the natural world have been so effectively silenced and disenfranchised. In Africa, however, the clash of cultures is still evident and it is difficult not to see its national parks as embodying sets of western, indeed colonial, values that are still at odds with those of many Africans themselves.

It would be a mistake, though, to think that value systems readily locate themselves along broad cultural, linguistic, and national lines. First, this suggests too little a role for shared values, either deriving from a shared human nature, or emerging, through a shared commitment to reason, in the course of dialogue and debate. And at the same time it downplays the extent to which seemingly entrenched cultural divisions are bound up with economic considerations – as we have noted, it's easier to advocate a respect for nature when the costs are not felt in your stomach. Second, the broad-brush approach to values ignores, too, just how much disagreement there can be within the one cultural and political frame. As the changing face of the Yosemite valley shows, claims to value nature, even among those professing a western, liberal, democratic outlook, can vary widely in their interpretation, from those who see nature as merely the backdrop to recreation, to those believing it should exist for its own sake.

The African example is illustrative, too, of a further point that needs to be made here. In the tables above I made some comment on the purposes for which parks exist. But these are the stated purposes, and it is unwise to assume that the real and stated purposes coincide. Though animal conservation is high on the publicized agenda, many environmentalists and ecologists are more concerned about the preservation of less charismatic species and ecosystems, focusing on the higher mammals more for promotional and fund-raising reasons than because of beliefs as to their true worth. We should take little of what anyone says about environmental issues at face value.

I want to end this section more or less where I began it, with the kind of Wordsworthian attitude to nature that, I suggested, lies somewhere behind the formation of national parks both in this country and in many places abroad. What that attitude suggests, perhaps, is that there may be a third way between two approaches mentioned above, the purely recreational on the one hand, and that which sees nature as having its own value, on the other. For what Wordsworth, along with Ruskin, Emerson, Thoreau and Muir, believed was that the experience of nature can benefit the human spirit, can help make us, in some sense, psychologically whole and help us realize our potential. This often connects with, but is still different from, the view that nature is beautiful, both because we can be moved by things that are not simply or merely beautiful, and because making this spiritual connection brings out more clearly than does the aesthetic concern the links between the human and the non-human worlds.

To claim in this way that nature, and in particular the large-scale nature of a national park, can meet certain of our spiritual needs, is, once again, to claim for nature some sort of objective value. The claim is not merely, and uncontrover-sially, that some people like to get away from it all, but that in a certain sense this relief from or antidote to culture, civilization and the city is of benefit to all, or at least to nearly all of us. Is it also to claim that nature, in affording us this benefit, is of instrumental value? That isn't quite right: first to experience nature isn't to use it and, second, even if there are benefits to be had they are not ones that you can simply choose to pursue. Such benefits will remain elusive so long as you think of nature merely as a means to their acquisition.

Summary

Although it is often tempting to privilege looking where landscapes are concerned, a wider and perhaps deeper experience of nature involves much more than this. And it is no accident that the impetus to the creation of national parks develops at around the time that an awareness of and urge for this deeper response to nature becomes widespread (around the mid- and late-nineteenth century). Even today, the kind of connection with nature offered us within many areas of the different national parks has undeniable spiritual value.

Yet there is more to say here. At least in part national parks are intended to offer scope for recreation, health, fresh air for human activity on a less elevated plane. Increasingly they are often seen as valuable preserves for non-human nature, as sanctuaries where birds, trees, insects, as well as the more charismatic mammals have some hope of survival. So understood, national parks are less well disposed to human activity. Different visions of what the park is for inevitably conflict, with recreational and ecological concerns perhaps most clearly at odds. And internal aims for the park conflict, too, with outside interests. Who wins? If parks are to survive then purists' visions of their aims and purposes will need to be abandoned, or at least disguised.

4   Wilderness

The landscape and parks of England are almost completely without wilderness. Virtually everything you see is the result of human management, or mismanagement of the land. If we want true wilderness, we have to look elsewhere.

Yet as soon as we think about where to look, we find we need to ask ourselves, just what is wilderness? This isn't as straightforward or as innocent a question as it might seem. It isn't straightforward, for none of the likely answers is easy to defend against its rivals, and it isn't innocent because, to many people's thinking, there is some sort of special value attached to wilderness. To describe a place as a wilderness is already to suggest that it is in some way significant and worth preserving.

4.1 What wilderness is

Wilderness, as any dictionary will suggest, is in some sense a wild, uncultivated, uninhabited region. But can we be more precise?

One suggestion, a place wholly unaffected by people, implies that there are no wilderness areas anywhere on the planet (see Figure 2.23). For everywhere is affected by people, if only because of the global result of our affect on climate and then its affects elsewhere. As there surely are some wilderness areas remaining even today, this suggestion has to be rejected.

Figure 2.23 The surface of the moon: is it still a wilderness?

Another suggestion, an area wholly devoid of human habitation, looks to fare better: for there are such areas – parts of some deserts and mountains, much of Antarctica, almost all of the bottom of the sea – still to be found. But why are people so important here? Is it really true that a handful of scientists living in a research base can compromise Antarctica's wilderness status? Similar questions can be asked about Alaska, parts of Africa, and the Australian outback. Why should we think that one area, lacking human inhabitants, is a wilderness, while a very similar area, but now sparsely populated, is not? Human beings can, of course, make a very big difference to a place, but they needn't always do so.

A third suggestion distinguishes between different kinds of human activity. So long as human beings leave the natural world more or less as it is, then wilderness survives. But when they aim to and succeed in reshaping their surroundings, when they 'subdue' or 'control' nature, then the wilderness is no more. Thus agriculture is antithetical to wilderness, while the nomadic life, that of the hunter-gatherer, can exist alongside it.

This way of understanding wilderness best fits the distinctions we want to draw. For many people (anthropologists and travel agents among them) the mere fact of human habitation doesn't prevent an area from being considered a wilderness. But there isn't wilderness in England, not because the whole of the countryside is heavily populated, but because even seemingly wild and natural places are shaped by centuries of agriculture, clearance and woodland management. Even so, there can be no firm lines here. Human cultures change not so much by step as by degree, with pastoralism, as practised for example by the Maasai, falling between the hunter-gatherer and the settled agricultural existence (**Reddish, 2003**). Is there wilderness still in East Africa? In a sense, we can say what we like about such a borderline case.

4.2 Why wilderness matters

Allow that there is no sharp line dividing wilderness from non-wilderness areas, and that wildness is matter of degree. Allow too that there isn't good reason to think that wildness is fatally compromised as soon as human beings attempt to live within it. Important questions about wilderness can now be asked.

Activity 2.4

In what ways might it be said that wilderness is valuable? Thinking again about instrumental and non-instrumental values will get you started. You should be able to think of three or four such ways, even if you don't find them all equally convincing.

Comment

Wilderness has, for many, a straightforward instrumental value. Large tracts of uninhabited or near-uninhabited land hold out the offer of untold resources, which can be harvested away from the public gaze and free from public interference. Increasingly, many governments want to exploit the uninhabited and sparsely inhabited parts of the world (such as Antarctica, Alaska, Siberia, the ocean floor) for oil, gas, minerals, etc. Many scientists see wilderness as instrumentally valuable in a different way, either functioning as a giant laboratory, or as an environment where rare and valuable species are found. Wilderness holds different sorts of values for those of us who, like Thoreau or Muir, feel from time to time the need to escape from the pressures of industrial society and to find there some kind of inner peace. Wilderness here is seen as a repository of spiritual values. Whether or not wilderness is aesthetically valuable will vary from case to case. While some wilderness areas will be widely seen as beautiful, a taste for others will be rather more a minority affair. Whereas some sort of attempt is often made to enhance or preserve beauty within national parks (see Figure 2.24), this isn't, and cannot be, high on the

Figure 2.24 Painted rocks in Morocco. Aesthetic values and wilderness values in collision? (Photograph by Christina Berlucci.)

agenda of wilderness protection. Finally, some people think the existence of wilderness is valuable just in itself. They think it matters that there are pristine wilderness areas left on this planet, and that the importance of wilderness is independent of its instrumental, spiritual or aesthetic values.

4.3 Changing values

As with landscape, our attitudes to wilderness have, over the centuries, undergone a process of substantial change. Contrasts made in the Bible, between the paradise of the carefully nurtured Garden of Eden, and the barren and desolate wilderness beyond, shaped western attitudes to the natural world for hundreds of years. Puritan British settlers in New England were hostile to and fearful of the countryside around them, and thought it a 'hideous and desolate wilderness', or, in the words of a poet among them:

> A waste and howling wilderness
> Where none inhabited
> But hellish fiends and brutish men
> That devils worshiped

> (Michael Wigglesworth, 1622; quoted in Sagoff, 1988, p.125)

Yet it wasn't long before the seemingly boundless frontier lands were seen less as a threat and more of a promise. Partly due to the influence of the philosopher and political theorist John Locke, wilderness was increasingly viewed as a valuable commodity, one that we had a moral duty to exploit. But, roughly a century later, this mercantilist response to wilderness started to be challenged by those who found town and city life increasingly industrialized, increasingly governed by bourgeois values and antithetical to what seemed the proper development of the human spirit. Rousseau's response to the mountain wilderness in his native Switzerland, like that of the American transcendentalists, Thoreau and Emerson, is linked here with the attitudes of English Romantics, in which wilderness is seen as a model of innocence and purity (see Figure 2.25).

> Why should we not have ... our national preserves ... in which the bear and the panther, and some even of the hunter race, may still exist and not be 'civilized off the face of the earth' – our forests ... not for idle sport or food, but for inspiration and for our own true recreation?

> (Henry David Thoreau, 1858; quoted in des Jardins, 1993, p.173–4)

This view, too, has met with criticism. Many in the twentieth century have accused the Romantic view of making a number of mistakes. Such a view is considered, by those advocating a more analytical approach, both to under-estimate human involvement in so-called wilderness areas, and to overestimate the purity and general benignity of non-human nature. The Romantics are also thought mistakenly to believe that, were it not for our involvement, nature would

Figure 2.25 *The American Wilderness*, 1864, 64.1 by 101.6cm, Asher Brown Durand.

change slowly, if at all, and (perhaps most importantly) to follow in religion's footsteps in assuming there is a firm and morally significant divide between the human and non-human realms.

Current attitudes to wilderness reflect different aspects of this complex history. While some are unnerved by pristine nature, others feel themselves more at home there than in towns and cities. And while some think the value of wilderness lies in what use we can make of it, others insist it has value in itself, and shouldn't be viewed simply as a resource for human preferences and needs. This mix of attitudes and values helps explain why disagreement as to the future of wilderness can be so profound. If you value wilderness for its resources, or because it contains important plants or animals, or for tourism, or because it is beautiful, then you'll have no strong objection to its management or modification in order to maximize the value it contains. If, on the other hand you value, as do many scientists, its pristine condition, or you think that our occasional encounters with it are of spiritual value, or even that it is valuable just in itself, then you'll urge that wilderness be preserved. We've discussed many of these values at sufficient length earlier. What I'll focus on for the remainder of this chapter is the last of those mentioned here, the belief that wilderness in particular, or nature more generally, is valuable just in itself.

4.4 Deep ecology, wilderness and intrinsic value

deep ecology

Deep ecology is a relatively recent, and still fashionable movement, which, though professedly worldwide, is strongest in those developed countries that might still claim to include aspects of wilderness areas – America, (and particularly California and the west), Australia and Scandinavia. Perhaps its best known advocate is the Norwegian philosopher, Arne Naess, who in a long career has shifted from a narrow to a much broader understanding of how philosophy is to be pursued. But it takes in non-philosophers as well, including poets, scientists, historians and political activists. Deep ecology is a radical movement, one that criticizes both traditional philosophy and traditional environmentalism for being hierarchical, inclined to put human beings, and human reason above all else. And it is influential. Its ideas (see Box 2.3) have helped shape the thinking of many well-known environmentalist groups, including EarthFirst!, Greenpeace, Friends of the Earth, road and runway protesters, and many of those objecting to what they see as the interminable spread of global capitalism.

Box 2.3   Deep ecology's basic principles

1   The well-being and flourishing of human and non-human life on Earth have value in themselves (synonyms: intrinsic value, inherent value). These values are independent of the usefulness of the non-human world for human purposes.

2   Richness and diversity of life forms contribute to the realization of these values and are also values in themselves.

3   Humans have no right to reduce this richness and diversity except to satisfy *vital* needs.

4   The flourishing of human life and cultures is compatible with a substantial decrease of the human population. The flourishing of non-human life requires such a decrease.

5   Present human interference with the non-human world is excessive, and the situation is rapidly worsening.

6   Policies must therefore be changed. These policies affect basic economic, technological and ideological structures. The resulting state of affairs will be deeply different from the present.

7   The ideological change is mainly that of appreciating *life quality* (dwelling in situations of inherent value) rather than adhering to an increasingly higher standard of living. There will be a profound awareness of the difference between big and great.

8   Those who subscribe to the foregoing points have an obligation directly or indirectly to try to implement the necessary changes.

(Devall and Sessions, 1985, p.70)

Deep ecologists claim to take the long view. Following arguments set out by the historian Lynn White, they see the origins of the present environmental crisis in the beliefs and attitudes of the Jews, Christians and Greeks. For so-called western civilization has, from the start, set itself in opposition to nature, emphasizing the differences between human beings and the rest of creation, and insisting on our duty to organize and control the natural world, to have dominion over it. And deep ecologists believe that many recent attempts to remedy our environmental problems don't look far enough into these root causes, and so adopt a sticking plaster approach to a range of ills that demand addressing at a deeper level. Thus animal rights and animal welfare concerns, a focus on the aesthetics of nature and a belief in conventional political solutions all represent, or so it is alleged, shallow, human-centred approaches to the natural world.

At the centre of deep ecology's critique is the notion of **intrinsic value**. Its proponents insist that the environment, and what it contains, is not simply a tool, a set of resources that can be drawn upon as and when we choose, conserved perhaps for future generations, but not at all valued for its own sake. Nature, whatever utilitarians may say, is non-instrumentally valuable.

But is there anything radical or surprising in this? Isn't it perfectly clear that the natural world isn't valuable merely instrumentally, as a resource? It is obviously, and often rightly, valued for non-instrumental reasons, for its aesthetic and spiritual properties. So isn't the deep ecologist's case already made?

In fact it isn't. While acknowledging that health, pleasure, beauty and the like are genuinely valuable, deep ecologists insist that if this is as far as we go, our position is still incomplete, still shallow. It is still too tied to human beings and their narrow and one-sided slant on the world. It is still too anthropocentric. For the deep ecologist's idea of intrinsic value is of something that is valuable in itself, quite irrespective of human concerns, or of a distinctly human perspective. When they believe, for example, that it would be a bad thing if tigers became extinct, they don't think this is bad because of the position of tigers in the food chain, or because they have magical properties, or because we find them beautiful. They think that it would just be a bad thing, a loss of value, were these magnificent creatures to disappear forever. 'Magnificent', though, is still misleading, for that, according to deep ecology is again to look at, and to value tigers from the human point of view. Take some humble beetle, one that looks as nondescript as the next, and which you might even tread on without a moment's thought (see Figure 2.26). Deep ecologists want to say its extinction would be a bad thing too, for it, just like the tiger, is intrinsically valuable, valuable in and of itself. Similarly for landscapes. We might value the preservation of some landscape – say, the high mountain ranges of Pakistan – because it is beautiful, or sublime, or because within such places we can find an antidote to the travails of modern life, but, according to deep ecologists, to single out these landscapes, and ignore others is again to evaluate the natural world in accordance with the way it meets certain human criteria.

Does deep ecology have it wrong?

Unsurprisingly, many people have been unpersuaded by the deep ecologist's position. Some objections come from the tradition that deep ecology wants to criticize. Others come from parts of the tradition it wants to praise. I'll focus first on the latter, and the question of anthropocentrism.

The Indian sociologist and historian Ramachandra Guha has argued that deep ecological positions are neither as broadly based, nor as cogently argued as its

Figure 2.26 An eighteenth-century microscopy plate of a beetle. Is such a creature valuable just in itself?

protagonists believe (Pojman, 2001). What he describes as an 'obsession with wilderness' derives, he insists, from what is in the end a parochial American perspective, and one that is ill-suited for export to parts of the world where human beings and the environment are not as easily separated as they have been in the USA. So Guha objects, for example, to the impetus behind *Project Tiger*, the system of parks that 'sharply posits the interests of the tiger against those poor peasants living in and around the reserves' (**Hinchliffe and Belshaw, 2003**). This impetus, or so he claims, comes first from international conservation agencies, mostly American backed, and second from members of the Indian feudal elite whose stance has shifted from that of hunter to conservationist while steadfastly ignoring the concerns of the poorer members of the community. The criticisms here echo those we've already raised in connection with the African parks – a concern for animals above people is symptomatic of Western values, values available to those who, because their economic position is secure, can indulge in a range of preferences and tastes. The attempt to impose such values beyond their narrow cultural setting is an example of the sort of neo-colonial activity that many of us, deep ecologists included, will claim now to be a thing of the past.

Nor, Guha insists, can deep ecology revive its global pretensions, and support its claims to offer solutions to worldwide problems via its frequent appeals to 'the wisdom of the East'. Although it makes much of the connection between its own environmentalism and that found in Buddhism, Taoism, Hinduism and the beliefs of, among others, native American and native Australian peoples, Guha finds this appropriation of selected parts of complex and often conflicting belief systems to be simplistic, patronizing and once again imperialistic in tone.

The deep ecologist's response

Deep ecologists are aware of these criticisms. They acknowledge that parts of them carry some weight. And they have offered a response:

One part of this response is important, welcome, and in the end, unsurprising. Deep ecologists increasingly insist that the human and the wild are not incompatible, and that wilderness areas can accommodate certain kinds and degrees of human habitation. They allow that they might in the past have appeared to put humans low on the agenda, almost to wish for the destruction of the human race, but now acknowledge this as an unfortunate but in some ways understandable distortion of their position. There is no reason to think this shift in position is other than sincere. For at the centre of deep ecology's rejection of what they see as the dominant western scheme of values, is their insistence that there can be no radical distinction between the human and non-human realms. Human beings are not different in kind from the rest of nature. So just as they cannot be secured a place at the top of the hierarchy, neither can they be pushed to the bottom.

Deep ecologists object, then, to the sort of activity that removes the Awhaneechee from Yosemite, just as they object to a worldwide ban on whaling, arguing that there is value in the continuation of a lifestyle, as in Norway, which has, as it long has had, whaling at its centre. Their shift in thinking here is reflected in policy changes in the African parks, where it is now becoming accepted that, as they have in the past, humans and animals can live alongside one another.

Yet even if deep ecology can successfully respond to the charge that it gives human beings too little space, where the central plank in their theory is concerned, that of intrinsic value, a rather more considered argument is needed. How, though, might this argument go?

Activity 2.5

Imagine that a wilderness area – say another island, as big as Zanzibar, but on the opposite side of the Indian Ocean – is newly discovered. Because of damage done in earlier cases, it's argued that this time human involvement must be ruled out completely, and forever. The island isn't to be set aside for scientists, or sensitive travellers. Its resources won't be held in reserve just until there's a pressing need to make use of them. And nor, when we've trashed the rest of the planet, can we use it to recharge our psychological batteries. Imagine, as well, that the island isn't particularly beautiful, or home to anything other than already well-known insects and plants. It's just a fairly ordinary wilderness. Could you be persuaded that it is important, or valuable, that this wilderness area remains? Suppose the human population somehow makes itself extinct and then, a thousand years later, the island is destroyed by volcanoes or earthquakes. There is now one wilderness less. Do you think that, if this were the island's fate, anything bad would have happened?

Comment

Notice that I've set up this example in such a way as to rule out potential confusion. Some people think it's bad when human beings destroy parts of nature. But if something is intrinsically valuable, then it doesn't matter how it is destroyed. Some people think that when destruction involves even non-human pain then a bad thing happens. But in this case there are no animals to be harmed. There aren't even human beings who might hear about the volcanoes, and feel sad about what's happened. It's pretty much a pure thought experiment, designed to get you to focus on the question of intrinsic value. If you think that the destruction of wilderness would be a bad thing, even when no one, and no thing is hurt in the process, then you probably think that wilderness is intrinsically valuable. If, however, you find it difficult to see what would be bad about this wilderness' disappearance then you are something of a sceptic about such value, and may well think wilderness is important simply because of its relation to human concerns. I myself doubt there is any such thing as intrinsic value. But of course you may think differently.

Summary

National parks often claim to be homes to wilderness. Yet, as we've seen, this notion is problematic. But though some people construe wilderness so narrowly that it becomes questionable whether there is any such thing, a better understanding, it is suggested, is one under which several such areas still exist.

But does wilderness matter? Should it be preserved? The focus here has been on the widespread claim that wilderness is intrinsically valuable. And my aim has been to suggest that while there is space for a range of values beyond the narrowly instrumental, the notion of intrinsic value is not one we should readily embrace.

5   Conclusion

The intention of this chapter has been to explore certain of the ways our beliefs are shaped by cultural inheritances of which we are not always fully aware. Notions of beauty and aesthetic value, our perception of and support for national parks and our approaches to wild nature and wilderness all have a complex history of cultural and social change behind them. They are also shaped by similarly complex geographies in the sense that our beliefs, values, and attitudes vary from culture to culture and place to place. We cannot answer questions like 'How are environmental understandings produced?' (whether those understandings are our own or those of others) or 'How and why do contests over environmental issues arise?' (whether those contests are about landscapes, parks and wilderness or other issues) without taking these temporal and spatial dimensions into account.

One response to this increased awareness of variation is to adopt a profound relativism, and to think that as all our values are flimsily structured on shifting sands, so there can be no legitimacy in our endeavours to defend them, to argue their case, to want their adoption elsewhere. Yet I've suggested that we can avoid this relativism – not only are commonalties discernible beneath the fractured surface (with some appreciation of landscape being a near-universal phenomenon) but there are good reasons for thinking that the promotion of certain values will serve us well, even if the truth in this is at some times, and in some places, obscured by an over-emphasis on material and narrowly utilitarian values.

The upshot of all this bears in fairly clear ways on the theme of this book. Conflict, be it about landscape, the use of parks, the nature and value of wilderness will, hopefully, frequently be resolved. For first, we might aspire to and in the end achieve a clearer and better grounded understanding of our own views, settling our internal inconsistencies and confusions; and second (though of course less frequently), we might succeed in persuading others that certain of our opinions trump their own. And thus, when these others are in conventional terms the more powerful, we might here look forward to the ascendancy of the weak over the strong. This may seem a too optimistic, even a vain hope. But to deny it is to deny that there's any proper space for dialogue, education and the process of politics.

References

Andrews, M. (1999) *Landscape and Western Art*, Oxford, Oxford University Press.

Appleton, J. (1996) *The Experience of Landscape,* Chichester, John Wiley & Sons.

Berry, G. and Beard, G. (1980) *The Lake District: A Century of Conservation* , Edinburgh, Bartholomew.

Blowers, A.T. and Hinchliffe, S.J. (eds) (2003) *Environmental Responses,* Chichester, John Wiley & Sons/The Open University (Book 4 in this series).

Clarke, K. (1949) *Landscape into Art*, London, John Murray.

Des Jardins, J.R. (1993) *Environmental Ethics: An Introduction to Environmental Philosophy*, Belmont, CA, Wadsworth.

Devall, B. and Sessions, G (1985) *Deep Ecology: Living as if Nature Mattered*, Layton, UT, Peregrine Smith Books.

Gill, S. (ed.) (1984) *William Wordsworth: The Oxford Authors*, Oxford, Oxford University Press.

Hinchliffe, S.J. and Belshaw, C. (2003) 'Who cares? Values, power and action in environmental contests' in Hinchliffe, S.J. et al. (eds).

Hinchliffe, S.J. and Blowers, A.T. (2003) 'Environmental responses: radioactive wastes and uncertainty' in Blowers, A.T. and Hinchliffe, S.J. (eds).

Hinchliffe, S.J., Blowers, A.T. and Freeland, J.R. (2003) (eds) *Understanding Environmental Issues*, Chichester, John Wiley & Sons/The Open University (Book 1 in this series).

Kemal, S. and Gaskell, I. (1993) *Landscape, Natural Beauty and the Arts*, Cambridge, Cambridge University Press.

Morris, R.M., Freeland, J.R., Hinchliffe, S.J. and Smith, S.G. (eds) (2003) *Changing Environments*, Chichester, John Wiley & Sons/The Open University (Book 2 in this series).

Muir, R. (1996) *Approaches to Landscape*, Basingstoke, Macmillan.

Opie, J. (1998) *Nature's Nation: An Environmental History of the United States*, Fort Worth, Harcourt Brace.

Pojman, L. (ed.) (2001) *Environmental Ethics*, Belmont, CA, Wadsworth.

Reddish, A. (2003) 'Dynamic Earth: human impacts' in Morris, R.M. et al. (eds).

Sagoff, M. (1988) *The Economy of the Earth: Philosophy, Law, and the Environment*, New York, Cambridge University Press.

Trillo, R. (ed.) (1999) *Rough Guide to Kenya*, London, Rough Guides.

Whelan, T. (ed.) (1991) *Nature Tourism: Managing for the Environment*, Washington, DC, Island Press.

Chapter Three

Power in the land: conflicts over energy and the environment

Andrew Blowers and Dave Elliott

Contents

1 Introduction: conflicts over energy projects

1.1 Energy: environmental impacts and technological choices

From the earliest times, human beings have burned wood to provide heat, to cook food and to frighten off predators. Subsequently, use has been made of the **fossil fuels** discovered in the ground – coal, oil and gas – along with hydroelectric power from rivers and, more recently, nuclear power, to create and run ever more complex societies, industrial processes, communication and transport systems. However, these attempts to extend our economic power by using natural resources have generated increasing problems. Most obvious has been the depletion of fossil fuels over the last one hundred years, during which time energy generation globally has expanded approximately tenfold.

fossil fuels

Burning of fossil fuels results in direct local impacts such as air pollution, although ways have been found to ameliorate these problems via emission control and clean-up techniques of various sorts. However, the main environmental impacts created by our combustion of fossil fuels now seem to be the planet-wide changes to the climate system. In order to deal with these we need to tackle the problem at source, and one way is to reduce emissions of carbon dioxide.

What are the options? We can reduce emissions to some extent by increasing efficiency at every stage. Firstly, the overall energy efficiency of power plants can be increased, for example by using gas turbines and by recycling some of the heat produced that is lost at present. Secondly, the potential for energy conservation through the use of more energy-efficient devices is very large: for example, just replacing all the light bulbs used in the UK with low-energy compact fluorescent lights would reduce demand by the equivalent of at least one large power plant; effectively insulating homes could save even more.

Even were these savings in energy to be made, however, it would still be necessary to make further significant reductions in emissions and this could be achieved by changing over to using non-fossil energy sources. That is not an easy option, since at present over 77 per cent of the energy used worldwide comes from fossil sources. Around 11 per cent comes from traditional biological sources (wood and dung), and only small amounts come from hydro-electric plants and nuclear power plants, and, so far, an even smaller contribution from new renewable energy sources such as wind power: see Box 3.1 and Figure 3.1.

Box 3.1 Energy around the world

Figure 3.1 gives an overview of the world's primary energy use at the beginning of the century in terms of basic energy content of all the fuels used around the world.

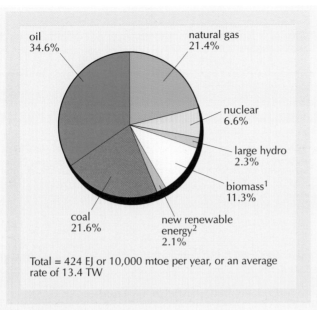

oil
34.6%

natural gas
21.4%

nuclear
6.6%

large hydro
2.3%

biomass[1]
11.3%

new renewable
energy[2]
2.1%

coal
21.6%

Total = 424 EJ or 10,000 mtoe per year, or an average
rate of 13.4 TW

Figure 3.1 Global primary energy
Notes:[1] Fuelwood, agricultural wastes and animal dung.
[2] Modern biomass, wind power, solar, geothermal and small hydropower.
EJ = exajoule = 10^{18}J; mtoe = million tonnes of oil equivalent = 10^{12} grams; TW
= terawatt = 10^{12} W. (See **Reddish, 2003.**)

Obviously the pattern of energy use for each individual country differs. For example, Norway and Brazil obtain more than half their electricity from hydro, France more than 70% from nuclear power, whereas in some developing countries there is very little use of electricity from any source. The UK gets around 25% of its electricity from nuclear power plants, nearly 30% from coal-fired plants and over 40% from gas-fired plants. Hydro and new renewables so far produce only about 3% of the UK's electricity.

Electricity is not the only type of energy used: in fact in the UK, for example, it represents only about 20% of the total energy used. In addition to generating electricity, gas is used for heating (representing about 30% of total UK energy used), and transport fuels account for about 40% of total energy. Coal and other types of solid and liquid fuel, for heating and industrial use, make up the rest.

In overall terms the industrialized countries comprise 20% of the world population but use about 60% of the world's energy. That translates into an even larger imbalance in *per capita* terms: the richest 1 billion people use 5 times more energy than the poorest 2 billion. However, some parts of the developing world are beginning to catch up in terms of *total* national energy consumed per annum. Certainly, in terms of overall energy use globally, the general trend is upward, increasing by approximately 1–2% each year.

Activity 3.1

According to Figure 3.1, nuclear power supplies only 6.6 per cent of the world's energy and yet Box 3.1 states that nuclear power plants supply 25 per cent of the UK's electricity. Does this mean that the UK produces much more nuclear power, proportionately, than the world average?

You will find the answer to this activity at the end of the chapter.

It is to the non-fossil sources that we must turn if we are to reduce emissions significantly. However, they are not without problems, notably the local environmental impacts they create: for instance and, as the next chapter will show, concerns are increasingly being raised about the social and environmental impacts of large hydro dam projects. In this chapter we shall look at the environmental costs to the local community of two of the possible non-fossil fuel sources – wind power and nuclear energy. They are examples of what have been termed **locally unwanted land uses (or LULUs)**.

locally unwanted land uses (or LULUs)

In approaching these examples we shall try to answer two questions. The first, derived from a key question for this book (set out in the Introduction), is:

> How and why do conflicts over the location of wind farms and nuclear waste facilities arise?

We shall explore this question by comparing these two technologies in terms of the different *value* positions inherent in local contests over land use and in relation to environmental impacts. The second question asks:

> What influence do changing power relations have on the outcomes of these conflicts?

This question requires us to focus on how the uneven distribution of political and economic *power* both shapes conflicts and influences the course of *action*. In doing so, we may gain some interesting, and perhaps surprising, insights into the power relationships between local communities, on the one hand, and broader economic, social and political processes, on the other.

1.2 The context of conflict

Wind energy and nuclear power present contrasting environmental problems. Windmills represent the revival of an older technology and, like watermills and tidal mills, generate energy from continuous natural energy *flows*. The advantage is that these flows – the winds, the waves, the tides – go on forever; they are naturally renewed. They are called **renewable** sources – to distinguish them from the non-renewable sources like the *stocks* of fuel in the ground which, whether fossil or nuclear, are inevitably limited and will run out. An estimate by

renewables

Shell International (1995) has suggested that by around 2060 approximately half the world's energy needs could be met by the use of existing renewable sources, such as hydro, and the so-called 'new' renewables, such as solar, wind, wave and tidal power, as well as the modern biomass technologies, including specially grown energy crops. Sugar cane and oilseed rape, for example, offer gas and diesel equivalents and fast-growing willow can provide wood chips to fuel power stations.

Activity 3.2

What disadvantages do you think there might be to the use of renewable energy sources such as the sun and the wind?

You will find the answer to this activity at the end of the chapter.

Using natural sources in some cases presents problems with intermittence. However, there is some symmetry in this approach to dealing with climate change. For, in effect, we have come full circle. We are trying to deal with climate impacts by using the natural flows created by the climate system. That can have advantages. For example, wind power is most available in winter when it is cold and the 'wind chill' factor is also highest. So wind power availability and the wind chill effect are directly correlated. Indeed conventional fossil or nuclear power plants have to be run up to maximum output to compensate for extra power demands due to wind chill; when using wind power, this compensation effect happens automatically.

However, not all the interactions between the use of renewable sources and the environment are positive. Using natural sources like the wind avoids the creation of air pollution, but there can still be local impacts such as noise pollution and the disruption of treasured rural views. Although these local impacts may be very small per kilowatt generated, compared with the global impacts of using fossil fuels, they are immediate and visible, and so can lead to local objections.

Nuclear power is generated by using the intimate substructure of nature – the atomic nucleus – to provide energy by causing materials such as uranium to undergo nuclear fission and so produce heat. This heat is then used to raise steam to drive turbo generators, as in conventional fossil-fuel power stations. Certainly the operation of nuclear plants does not generate carbon dioxide. However, relying on nuclear power seems to create at least as many problems for us as it solves, chiefly because of the risk of releasing radioactive material into the environment. In particular, there is the problem of finding somewhere to store the radioactive wastes that nuclear plants produce.

nuclear power

Although neither renewable projects like wind farms nor nuclear power plants produce any carbon dioxide directly, some *will* be generated in the process of manufacturing the materials and equipment for these power plants, since that involves the use of energy – and, for the moment, most of this will come from fossil fuel sources. It is therefore instructive to look at the 'energy debts' of these two technologies. A study of energy payback times has indicated that, over their full lifetime, wind turbines typically generate around 80 times more power than was used in their construction and operation; by comparison the figure for nuclear power plants is 16 times.

Aside from technical issues like this, and disputes over questions of economics and national energy policy, energy projects like wind farms, nuclear power stations or radioactive waste facilities provoke *local* conflicts which are related to, and influenced by, conflicts of interest, values and resources within society. To put it very simply, people may object to having major projects imposed on them mainly to service the energy needs of a wider population. There are also conflicts inherent in the fact that energy projects are expensive and are usually developed by large companies, owned and controlled by outsiders, with all the environmental and social cost, but few of the economic benefits, falling on local residents. These conflicts are often portrayed as straightforward examples of **NIMBY (not in my backyard)** reactions to the imposition of LULUs. But, they are often more complex, and involve a variety of economic (jobs, investment), environmental (aesthetic, pollution, risk), social (community) and political (local, national) issues.

NIMBY (not in my backyard)

In what follows, we shall interpret these conflicts and their outcomes by developing the analytical concept of power (see Chapter One and **Hinchliffe and Belshaw, 2003**). We shall identify the *power resources* that are deployed in these conflicts over land use and how these affect the power relations that determine the location of LULUs such as wind farms or nuclear waste facilities. In doing so, we shall be able to address our first question: how and why do conflicts over the location of wind farms and nuclear waste facilities arise? The case studies also reflect the changing nature of power relations and how *power as discourse* operates to promote, reinforce or challenge the uneven social and geographical pattern of energy projects. This will enable us to reflect on our second question: what influence do changing power relations have on the outcome of these conflicts?

2   Down on the wind farm

The case study which follows should help you to explore the motives and values of those objecting to wind farms, and to see how they have attempted to take action to overcome the sense of powerlessness often felt in the face of projects imposed on communities by outsiders.

Activity 3.3

As you read the case study on wind power, ask yourself:

> Were their reactions simply a variant of conflicts over rural versus urban interests? Did hostility to outsiders play a part?
> What about altruistic environmental protection and preservation concerns? Or were other factors at work?

Specifically in terms of power you should ask:

> What power resources could they bring to bear to pursue their objections?
> Is it possible to discern changes in power relations which favoured one side or another at different times?

Also, you will inevitably want to ask:

> Are the objectors right – are wind farms a bad idea?

As you read through this case study, you will form (or maybe change) your own view.

2.1 Introducing wind power

The use of wind turbines is seen as one way of generating power without creating dangerous pollutants and greenhouse gases. Around the world, wind power is the fastest-growing new energy technology, having jumped from more or less a zero contribution in 1970 to around 24,000 megawatts (MW) globally by 2002. (For comparison, a typical modern fossil or nuclear plant is rated at around 1000 MW: one megawatt is 1000 kilowatts, and 1 kilowatt (kW) is the rated power of a typical one-bar electric fire.)

The current designs of wind turbine have full power ratings of between 1 and 2 MW. Since the wind does not blow all the time, they will not, of course, deliver that power continuously. In practice, because of variability of the wind, wind turbines typically, on average, deliver around 35 per cent of their full power rating. Given that conventional power plants can deliver around 70 per cent of their full power rating, you would need roughly twice as many wind turbines to deliver the same amount of total power as a conventional plant. Thus to match the output of a large 1000 MW coal or nuclear plant you would need 1000 x 2 MW wind turbines (or, of course, fewer if they were larger machines). Large machines are much more efficient since the power output is proportionate to the *square* of the blade length. The power in the wind is also proportionate to the *cube* of the wind speed, so that selecting windy uphill sites has very significant efficiency benefits.

The UK has one of the world's best wind resources. The government's target, adopted in 1999, is to obtain 10 per cent of the UK's electricity from renewables by

(a)

(b)

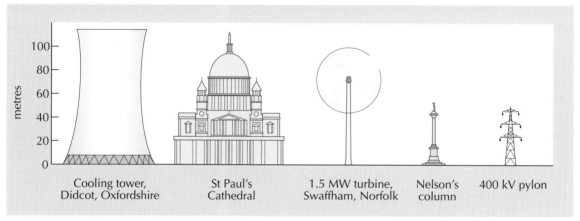

(c)

Figure 3.2 Wind power: old and new.
(a) There were once tens of thousands of traditional corn-grinding windmills around Europe.
(b) Modern wind turbines generate electricity and are much larger and more powerful. This modern wind turbine at West Somerton in Norfolk can generate up to 1.5 MW at full power, generating up to 4.5 GWh per year; the hub height is 67m and the rotor diameter is 66m.
(c) To give an idea of scale, this diagram shows a typical modern large wind turbine set against some familiar existing landmarks.

2010, with wind power, which is the most developed of the new renewable technologies, providing perhaps half of this. The British Wind Energy Association has suggested that this could imply the siting of around 2,500 wind turbines on land and about the same number offshore, based on multi-megawatt machines. By 2002 nearly 1000 turbines had been installed around the UK, mostly smaller machines, representing in total nearly 500 MW of generating capacity at full power: see Figure 3.3.

Figure 3.3 Sites of wind farms in the UK, 2002.
Source: British Wind Energy Association

2.2 Wind power in the UK: the rise of opposition

The UK wind farm programme, which got under way from 1990 onwards, owes its existence primarily to the 'non fossil fuel obligation' (NFFO), a cross-subsidy scheme introduced following the privatization of the electricity supply industry in that year. In effect it involved a small surcharge (around 1 per cent) on consumers' electricity bills to support renewable energy projects, and was part of a larger surcharge for supporting nuclear power, the total coming to around an extra 10 per cent on consumers' bills: this raised around £1.2 billion per year. The renewable projects were brought forward in a series of competitive rounds: NFFO–1, the first round, was opened in 1990, and NFFO–2 in 1991. These together included 58 wind projects totalling around 110 MW of wind turbine generation capacity.

Figure 3.4 These 400 kW wind turbines are part of the first wind farm in Britain at Delabole in Cornwall. The farm consists of 10 Danish-built turbines which together produce 4 MW of electricity at full power.

Overall, the public reaction to wind power was positive: see Table 3.1. Moreover, the level of support actually increased when people had experience of wind farms. For example, local opinion was sampled both before and six months after a 10-turbine scheme was installed at Delabole in Cornwall – the UK's first wind farm (ETSU, 1993): see Figure 3.4.

Table 3.1   Summary of public opinion studies around UK wind farms, July 1996 (percentages)

Location	Survey time	Sponsor/ organizer	Date	In favour	Against	Don't know
Delabole, Cornwall/ Devon	S & P	DTI	1992/93	84	4	11
Cemmaes, Powys	S & P	DTI	1992/93	86	1	13
Llandinam, Powys	P	CCW	1992/93	83	3	14
		BBC/UW	1994	76	17	8
Llangwyryion, Dyfed	P	CCW	1992/93	78	8	14
Rhyd y Groes, Anglesey	P	CCW	1992/93	74	17	9
		BBC/UW	1994	61	32	7
Taff Ely, Rhondda	P	BBC/UW	1994	74	9	17
Kirkby Moor, Cumbria	P	NWP	1994	82	9	9
Bryn Titli, Powys	S	NWP	1993	68	14	19
	P	NWP (Open day)	1994	94	3	3
Trysglwyn, Anglesey	P	NWP (Open day)	1996	80	11	8

Notes: NWP = National Wind Power Ltd (wind farm developer); CCW = Countryside Council for Wales (statutory body); BBC/UW = BBC (Wales) + University of Wales
Survey: Pre-construction = S; Post-construction = P

Source: British Wind Energy Association.

Particular impacts were considered in some detail. More than 40 per cent had initially thought visual intrusion might be a problem but this fell to 2 per cent subsequently. Concerning noise, initially only 14 per cent had expected this *not* to be a problem, but in the event this figure rose to around 80 per cent.

However, as the UK programme got under way, local opposition to some projects became increasingly significant. The response from environmental groups was mixed. Friends of the Earth, locally and nationally, maintained their long-held support for and promotion of wind power, as did Greenpeace and the Labour Party's environmental campaign group, SERA. The Royal Society for the Protection of Birds (RSPB) and the World Wide Fund for Nature (WWF) also supported wind power. Several other national conservation groups, however, have been critical of wind farms to varying degrees. The National Trust, for example, has expressed concern about inappropriate siting and the Ramblers Association has come out against wind farms. In addition, many of the regional conservation organizations made critical comments on location. The Countryside Council for Wales (CCW) argued that there 'should be a presumption against wind turbine development in areas of close proximity to sites with the benefit of statutory landscape designation status' (CCW, 1992). The Council for the Protection of Rural England (CPRE) called for greater scrutiny of projects (CPRE, 1991) and Tony Burton from the CPRE told *The Guardian* (11 March 1994) that, while they were not opposed to wind power in principle, 'the system of subsidies is putting pressure to build wind farms in quite inappropriate places, remote landscapes that have been protected for decades'. The Northern Devon Group of the CPRE saw the programme as a 'mad scramble' for lucrative hilltop sites, with wind turbines threatening 'to stride across the countryside of North and West Devon like a plague of triffids' (Allen, 1991). Barry Long from the CCW told *The Times* (21 August 1991), 'It will be hard to stand on a hilltop in mid Wales without seeing a windmill.'

During 1993 there were local objections voiced around the country, for example, in Devon, Cornwall and Yorkshire (no projects were initiated in Scotland until later), but perhaps the greatest number emerged in Wales. The UK's largest wind farm so far, at Llandinam, proved to be something of a turning point in the debate. There had been objections to its scale, on visual intrusion grounds, but in the event it was noise that proved to be the major problem. Several local residents claimed to be suffering from major disturbance, and there did indeed seem to be a significant problem for some residents in the valley below the ridge on which the 103 Mitsubishi machines are sited (Walker, 1993): see Figure 3.5. The local experience with this project stimulated significant objections to subsequent projects in Wales. An extensive debate ensued in the local press, with opinions often becoming polarized. Some colourful rhetorical allusions emerged. National Wind Power's wind farm at Llangwyrfon was alleged to sound like 'an old wheelbarrow being pushed along continuously', while their Cold Northcott project in Cornwall was described as sounding like 'a huge washing machine gathering speed to spin dry' (NATTA, 1993).

Figure 3.5 Some of the 103 wind turbines, each 35m high, that comprise a wind farm at Llandinam in Wales, one of the largest in Europe, with a capacity of 30.9 MW.

The amount of noise varies. Some individual machines may have been particularly noisy during their run-in periods and, in some topographical situations, sound can be amplified by resonance effects. However, in most cases, visitors are usually surprised at how quiet wind farms are, just a slight blade swish even close up, together with occasional gear train rumble. But people's response to noise varies. For example, some people are very sensitive to low grade background noise, and cannot sleep with a refrigerator running in the same room. Certainly, once a noise starts to be annoying, it can be detected even at very low levels. Fortunately, wind turbine designers have developed a new generation of direct-drive, variable-speed wind turbines, which are less noisy since they do not have gear trains and the rotor speed is better matched to the wind speed.

There is less that developers can do about visual impacts, and this issue began to attract national media coverage, especially after the setting up, early in 1993, of a national anti-wind lobby group, 'Country Guardian', dedicated to opposing 'the desecration of the coasts and hills by wind farms'. Its vice president, Sir Bernard Ingham, famously described wind turbines as 'lavatory brushes in the sky'. Ingham's extensive media contacts, from his time as press secretary to Prime Minister Margaret Thatcher and as public relations adviser to British Nuclear Fuels, were clearly invaluable (Aubrey, 1993). As a consequence, a campaign against the proposal to site 44 turbines at Flaight Hill, near Hebden Bridge in West Yorkshire, attracted national media attention, and also led to the involvement of a number of celebrities including some notable literary figures – complaining about what they saw as an 'assault on our literary and artistic heritage', given that the wind farm would be in Brontë country.

In addition to visual intrusion, wider preservation and conservation issues have also been raised, for example in relation to potential impacts on wildlife and ecosystems (see Box 3.2). However, in many cases simple self-interested NIMBY-type responses predominate, for example in relation to alleged impacts on house prices.

Box 3.2 Environmental impacts

In addition to the issue of visual intrusion which is a human concern, wind projects might be expected to have some impacts on wildlife and the environment. The issue of bird strikes has not been seen as very significant in the UK (the Royal Society for the Protection of Birds has generally backed wind projects) although it has been raised in the USA. Studies have indicated that birds avoid moving objects like wind turbine blades and are much more likely to collide with stationary power grid cables, as happens regularly. However, care must be taken to avoid migration routes, where large numbers of birds are involved.

Most other animals seem indifferent to wind turbines. Cows and sheep graze right up to the tower bases. Indeed some seem to welcome them as providing shelter from wind in winter. Some damage is invariably done to the local ecosystem during construction, and by the foundations for the concrete bases of the turbine towers, but the wind turbines and their bases can be removed, if that ever becomes necessary.

Wind farms are not allowed in the statutorily defined Areas of Outstanding Beauty (AONBs) or Sites of Special Scientific Interest (SSIs).

Local opposition was also directed against the fact that some proposals were made by large national or, especially, international companies funded by money from overseas, with most of the equipment used being of foreign origin. In this context it is perhaps not surprising that local residents have sometimes expressed resentment, not just at having to accept local disruption in order to provide power mainly for 'townies', but also at invasion by remote companies, and domination by absentee owners and by the power of foreign economic interests. Thus one anti-wind-farm leaflet spoke of Wales being 'in the forefront of being covered in swathes of ugly turbines to line the pockets of foreigners and greedy owners' (NATTA, 1994).

Activity 3.4

What are the main reasons for the local opposition to wind farms?

The answer to this activity is given at the end of the chapter.

In pressing their opposition to wind farms, local interests were able to apply their power resources, in particular the power available to them through the media. The press have shown considerable interest in the local debate over wind farms: all of the major national broadsheets (*Times, Guardian, Independent, Telegraph, Observer, Financial Times*) have carried reports, the coverage increasing as objections mounted, with the emphasis mainly therefore being on the negative side. The broadcast media, nationally and locally, also generally adopted a fairly critical – and in some cases hostile – approach. The local press in the relevant areas carried regular news stories and features plus extensive letters from local people, with objectors usually dominating.

In response, the wind industry lobby group, the British Wind Energy Association, asserted that 'the controversy to date has largely revolved around misconceptions and misinformation distributed by groups aiming to stifle wind energy development completely' (Harper, 1994). Certainly there have been cases of misrepresentation. For example, the 10 per cent NFFO levy has sometimes been cited by wind farm objectors as being the extra cost imposed on electricity consumers by the wind farms. In fact, as we noted earlier, in 1990 the bulk of this 10 per cent supported nuclear power: only around 1 per cent went to renewables and only some of that to wind projects. (Since then a Renewables Obligation has been introduced which requires energy supply companies to source 10 per cent of their electricity from renewable suppliers by 2010, but limiting extra charges to consumers to 4–5 per cent by 2010.)

However, opposition cannot be written off as simply mistaken, or as just reflecting effective lobbying by a small minority of activists. Even though the opinion polls seem to indicate generally high levels of support, the lobbyists could not operate effectively unless there was opposition at the grassroots. While objections focused mainly on the amenity issues, broader concerns were also used to reinforce the case. Some objectors accepted the global argument in relation to climate change but claimed that wind farms could not contribute significantly, so that the local impact was not justified. Others favoured alternative energy options with energy conservation often being seen as a better choice. For some others, the central issue was what they saw as 'profiteering' by 'greedy developers', who make use of the 'extensive subsidies' without, allegedly, being concerned with the impact on the local environment. By using both local impacts and broader concerns the opponents to wind farms were able to build up a wider constituency of protest.

On the other side, strong support for wind farms was given as part of a positive commitment to the future with the aim of helping to avert the threat of global warming and avoiding the dangers of nuclear power. Many supporters also said that they actually liked the look of the wind farms. An attempt to set the debate in a more productive vein was made by the extensive review carried out by the all-party House of Commons Welsh Affairs Select Committee in 1993–94. This took evidence from all the protagonists. The Committee concluded that as long as they were sensibly planned, wind farms could be acceptable in Wales and, in effect, rejected some of the more aggressive claims of the objectors (Welsh Affairs Committee, 1994).

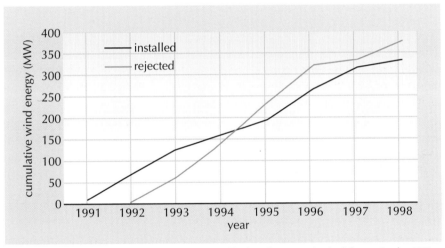

Figure 3.6 Cumulative wind energy of installed and rejected wind farms in the UK, 1991–98.
Source: G. Sinclair at Environmental Information Services.

Nevertheless, opposition continued. Indeed it actually became more effective, resulting in nearly 70 per cent of UK projects in recent years failing to get planning permission (see Figure 3.6). Between September 1991 and December 1993, 12 wind farm proposals went before planning inquiries and 9 of these were approved. But between January 1994 and January 1999, only 2 of the 18 proposals called in for decision by the Minister won approval following a planning inquiry. Though only a proportion of wind farm proposals are called in for planning inquiries the trend seems clear. Dr Peter Musgrove, from National Wind Power, told *The Times*, 'Since 1994, planners and inquiry inspectors have been giving progressively less weight to the clean energy benefits of wind farms and progressively more to their negative and subjective assessment of visual impact.'

As *The Times* (9 January 1999) put it, 'unless urgent action is taken, many firms will leave Britain for wind power opportunities overseas', and the UK would find it hard to meet the target of obtaining 10 per cent of its electricity from renewables by 2010.

2.3 Possible resolutions: local involvement

As we have seen, there is broad support for wind farms in the UK. In a national poll carried out for the RSPB in 2001, only 3 per cent of those asked were opposed to on-land wind farms. Looking at the local level, where there is direct experience of wind farms, support actually seems to grow once wind farms have been established. For example, a study of people living near Scotland's first four wind farms indicated that 40 per cent had felt there would be a problem but only 9 per cent reported any actual problems after the event. Overall, two-thirds of those asked found 'something they liked' about wind farms (NATTA, 2001). However, there is also some strong local opposition and, even if that is a minority, it seems to be slowing the rate of deployment dramatically. How might this situation be resolved?

We have to recognize that not all wind projects are necessarily well thought out, and there is a variety of practical steps that could be taken to improve the situation, such as better location of wind farms, and the use of waste land or old industrial sites. The introduction of more efficient, variable-speed wind turbines, which can operate effectively at lower wind speeds, could reduce pressure on upland sites. Costs have also dropped dramatically since the early days. Some wind projects have been able to go ahead at contract prices of around 2p/kWh, which is cheaper than most conventional sources of power, and one-third of the cost of wind projects established in 1990. That makes it more economically viable to choose less windy (and therefore possibly less visually dominant) sites.

However, to ensure that these technical and economic improvements translate into environmental/locational improvements will probably require something more coherent than the UK's current rather ad hoc approach to planning in relation to wind projects, which operates on a site-by-site basis. Although it is sensible to assess each project on its merits, with local issues and concerns being taken into account, this can lead to expensive planning delays. One solution would be to develop some form of zoning, with areas suitable for development identified in advance: this has been the approach taken in Denmark. Something like this is in hand in England with the development of a regionally based planning framework and local councils setting targets for the amount of renewable energy to be obtained in their regions (DETR, 2000; DEFRA, 2000). The government is also exploring how to speed up planning procedures. However, any weakening of planning control is likely to be seen as reprehensible by most environmental groups, even by those in favour of wind, since it might open the way for less desirable projects.

Meanwhile the government is considering expanding the renewable target to 20 per cent by 2020, with Scotland's devolved administration proposing its own target of 30, or even 40, per cent by 2020. Wind power would inevitably play an increasing role in programmes of this sort.

Nonetheless, local opposition is likely to persist in many places unless a way is found to address the feeling of powerlessness that some opponents evidently experience in the face of wind projects imposed on their communities. Most of the wind projects are owned and controlled, one way or another, by a few large companies, and, as we have seen, this has sometimes been another source of resentment. For example, one of the issues raised during the local campaign against the 39-turbine wind farm given planning permission for a site at Cefn Croes in mid Wales was that it would not benefit local people and was being 'imposed on the community by Enron Wind, a subsidiary of the shamed US multinational' (*The Guardian,* 20 February 2002).

By contrast, opposition to wind farms has been far less apparent in Denmark and Germany, where the majority of wind projects are owned by local people. The growth in privately owned as opposed to utility-owned projects in Denmark has been quite striking (see Figure 3.7). Around 80 per cent of the 3000 or so wind projects are owned by individual farmers or local wind co-operatives – the Wind Guilds – who share in the profits (Figure 3.8). Overall, more than 100,000 people own shares in local wind projects in Denmark, out of a population of 5 million. By 2002, 18 per cent of Denmark's electricity was generated by wind. Similarly, in Germany, local interests dominate ownership – and by 2002 Germany had installed around 10,000 MW of wind capacity, compared with the UK's 500 MW. Thus, while in the UK some people have been clamouring for wind projects to be

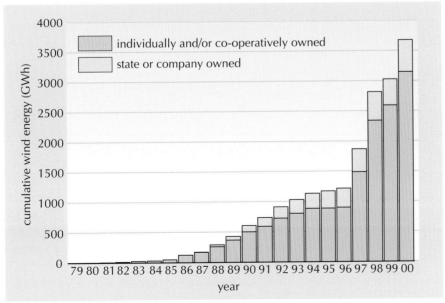

Figure 3.7 Growth of local ownership of wind projects in Denmark.
Source: Gipe, 1995, Figure 2.3, p.59, updated with data provided by Gipe.

Figure 3.8 A general meeting of the Middelgrunden Wind Turbine Co-operative in Denmark.

halted, in some other parts of Europe many people seem to be clamouring to be able to benefit economically.

If wind projects involve some form of local control and local economic benefits, the local impacts may be judged more favourably. Danish enthusiasts for local ownership often repeat the old Danish proverb – 'Your own pigs don't smell'. Put more positively, if it is people's sense of powerlessness that is at least part of the problem, then the solution is for them to develop their own power resources, by getting directly involved, via ownership. This can lead to direct economic benefits or to more indirect gains such as local economic regeneration. For example, farmers, who are finding it hard to survive economically by producing food, might diversify into wind power and other renewable projects, thus strengthening the local economy (Toke, 1999).

So far, there is only one wind co-operative in the UK – Baywind in Cumbria – although other initiatives are under way. One group in Wales has taken pains to consult widely with local people, and in 2001 it organized a public referendum in the area to decide whether a wind co-op project should go ahead: the result was 57.5 per cent in favour, 42.5 per cent against. However, even with conventional commercial wind projects, there are some local economic benefits, in terms of employment during construction, possible tourist attraction and rental income for local farmers for wind farm development. In some cases wind farm developers have offered packages of 'community support measures'. For example, National Wind Power provided £100,000 for a Charitable Trust to be used to support local schools, colleges, students, apprentices and training schemes in the area around its wind farm at Bryn Titli in mid Wales. Schemes like this might give local communities a vested interest in the project; failing that, they might at least be seen as compensation for any disbenefits; alternatively they might be seen, by opponents of the project, as a form of bribery. Another possibility might be to charge less for electricity to those who live near wind farms thereby compensating them for loss of amenity.

Activity 3.5

Can you foresee any problems in such schemes of compensation?

Compare your answer with the one given at the end of the chapter.

Another form of involvement, albeit rather diffuse, can occur if consumers contract with one of the dozen or more green power retail schemes that are on offer. In these, electricity suppliers promise to match the power consumers use with power bought in from renewable sources. In some schemes a small surcharge is made. By 2002 over 45,000 people had signed up in the UK, and the number is expected to grow significantly, as has happened elsewhere (in 2002 there were 280,000 green power subscribers in Germany, 680,000 in the Netherlands and 900,000 in Denmark). Some of the schemes use wind power, and participating consumers are able to identify the sites used. This may not be ownership, and it may not be a local plant, but it does make some sort of link between what you get out of the socket in the wall, and the source of the power. Moreover, in some schemes consumers actually donate a surcharge directly to a fund or trust which invests in specific renewable energy projects, usually locally: in these cases, there is an even more direct sense of local involvement.

2.4 Resisting wind power: what are the alternatives?

One of the fundamental issues that emerge from this case study relates to people's feelings of powerlessness to protect their interests and their lifestyles. The main concern that has been expressed is over visual intrusion, but this can reflect more than just an altruistic or personal concern for non-instrumental values. Opponents also express instrumental concerns, for example about the possibility that house prices might fall or, in some locations, about the potential loss of revenue from tourism. Where people feel powerless in relation to issues like this, one response is to try to create an alternative, defensive, power base, for example by organizing local campaigns and by seeking media support. However, in some cases, the development of environmental awareness can lead to some contradictions when it comes to choosing the focus for specific campaigns. For example, if you are concerned to protect your local environment, you can seemingly have far more effect by opposing a local wind farm than by campaigning to halt global climate change. The former can be seen as an immediate and visible threat; the latter is remote, longer-term and, in any case, ostensibly beyond your influence.

Some objectors have tried to adopt a more positive approach by pointing to alternative possibilities – for example, offshore wind farms. The energy resources offshore are very large, and the environmental impacts associated with tapping them are generally low. Offshore wind farms have already been developed off the coast of Denmark, The Netherlands and Sweden and sites for 18 have been proposed around the UK coast. Clearly this could be a useful option, but, given the UK's target of 10 per cent renewable energy by 2010, it is likely to be in addition to, rather than instead of, on-land siting. Moreover, offshore siting is more expensive than onshore wind, primarily due to the need to transmit power by undersea cable to the shore; also, unless offshore wind projects are sited well out to sea (thus increasing cost further), there could still be objections to visual intrusion. Indeed, there have already been strong local objections to some of the first offshore wind farms proposed in the UK, despite their location several kilometres off the coast: the project proposed for the Solway Firth, for example, is on a sandbank 8 kilometres from the shore.

Figure 3.9 Wind turbines located offshore at Middelgrunden in Denmark.

Energy conservation is obviously another alternative, but once the opportunities for cheap and easy savings have been exhausted, the cost-effectiveness of energy-saving measures is not much different, in terms of carbon emissions avoided per pound spent, than that obtained from wind power projects. Moreover, if the climate change problem is as serious as many think, and we wish to avoid the nuclear option, then we will need all the energy conservation and all the renewable sources we can reasonably muster. It is not a matter of either/or. This conclusion becomes more stark if we assume, as seems likely, that, even given a major commitment to energy conservation, overall demand for power will continue to rise.

2.5 Conclusion: power and action

Activity 3.6

At the beginning of this chapter we posed two key questions:

> How and why do conflicts over the location of wind farms arise?
> What influence do changing power relations have on the outcomes of these conflicts?

Think about your answers to these questions before reading on. You may also like to look back at the questions posed in Activity 3.3 (p.89) and think about them, too.

We can now consider our answer to the first key question: how and why do conflicts over the location of wind farms arise?

On the one hand are local (mainly rural) communities supported by some broader-based conservationist groups, opposed to wind farms predominantly on aesthetic grounds but who are also concerned about potential economic impacts on house prices or tourism. There is resentment, too, at the intrusion of outside interests developing a local resource while the benefits go elsewhere. On the other hand are the supporters and promoters of wind power including some environmental organizations, government bodies and wind power companies who view wind power as one of the more viable alternatives to fossil fuels or nuclear power. Although the balance of power seems to be tilted very much in favour of the wind lobby, the local communities and conservation groups have nevertheless often been able to win out against their opponents. Moreover, these successes have been achieved without extremism on either side. Promoters of wind power – government included – have so far chosen not to muster all the resources available to them, while their critics have avoided the direct action protests associated with, for example, opposition to road-building. Within the generally even-tempered terms of this debate, the anti-wind campaigns have in effect created a countervailing power base which has, in some cases, been able to resist the power of the developers and, to the extent that they support wind power, the government.

We have also found some answers to the second question: what influence do changing power relations have on the outcomes of these conflicts? At least part of the debate concerns the relative distribution of costs and benefits of new energy supply technologies such as wind turbines. Opponents regard wind farms as a cost to their community and seek to prevent them or divert them to other locations (especially offshore). In order to shift power relations in favour of wind projects it has become clear that rural people near wind farms could be compensated for any local costs – or could share in the benefits. In short, it may be that wind energy requires a new discourse of co-operation and consensus if it is to make its planned contribution to the UK's electricity supply.

In countries like Denmark and Germany such a discourse already exists: rural residents are now actively involved in generating and selling power. The locus of power – economic as well as electrical – thus seems to have changed. Unless measures are taken to bring about similar changes in the UK, it could be that we will see even more local opposition to wind power, and possibly to other renewable energy projects that impact on rural areas, such as growing energy crops (as discussed in Chapter 1). If this situation were to develop, it could, ironically, have some parallels with the one which emerged when the nuclear industry attempted to site nuclear waste repositories in rural areas – that of massive local opposition, as the next case study illustrates.

The example of conflict over wind energy raises the issue of the degree to which local concerns, especially minority concerns, can be accommodated within wider development programmes. On the one hand, there is the risk that parochial interests can block policies with national or international significance; on the other hand, there is the danger that powerful government agencies might steam-roller legitimate dissent.

In the next case study, we move on to a situation where this type of power battle actually led to a retreat by the authorities. The case study of nuclear waste provides us with some insights into how and why conflicts over the location of nuclear waste facilities arise.

3   Nuclear waste: the power of the powerless

3.1 The emergence of the problem

The environment of the Blackwater estuary in Essex is discussed in **Blowers and Smith (2003)** in the first book in the series.

On the southern side of the Blackwater estuary, near its mouth, is the Bradwell power station, one of nine Magnox stations (the name derives from the magnesium alloy cans that encase the nuclear fuel) that comprised the first generation of nuclear power stations in Britain: see Figures 3.10 and 3.11. Bradwell generated electricity for 40 years until its closure in 2002. The site, a wartime bomber landing strip, was selected for three reasons: it had an ample supply of sea water for cooling; it was relatively close to the major centres of demand in the south east of England; but, at the same time, it was regarded as a 'remote' location. The site met the requirement that '... the first stations, even though they will be of inherently safe design, will not be built in heavily built-up areas' (HMSO, 1955, p.9). In practice, for Bradwell and the other early stations, 'the "remote" siting criteria were never meant to make full use of geographical isolation as a safety measure but merely to avoid locations within 20 miles of large towns' (Openshaw, 1986, p.109). Even on this measure Bradwell's location must be regarded as pragmatic. Although the immediate surroundings on the south side of the estuary have a very small population, within eight kilometres there are several villages and the small town of West Mersea (7,000 people), while within 25 kilometres there are the substantial towns of Colchester (with over 100,000 people) and Clacton (see Figure 3.11).

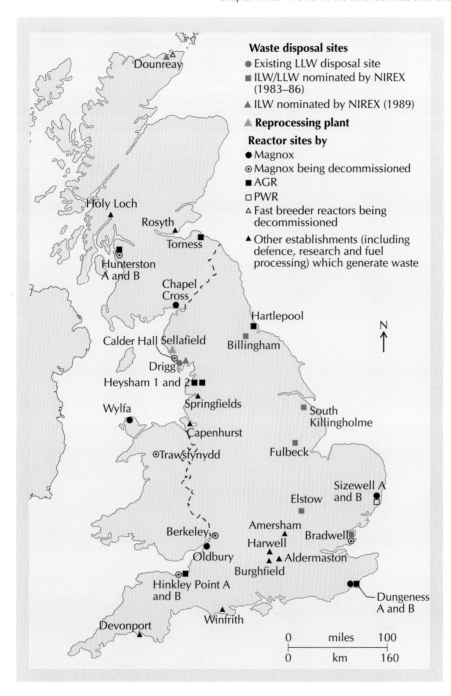

Figure 3.10 The nuclear industry in the United Kingdom.

The Bradwell nuclear plant was established with relatively little controversy in an era when nuclear power was an exciting new technology. But, in the 1980s, along with three other sites – at South Killinghome on Humberside, Fulbeck in Lincolnshire and Elstow in Bedfordshire, all in eastern England (shown on Figure 3.10) – it became a focus of conflict over proposals for a national repository for the shallow burial of nuclear waste (Figure 3.12). Although the proposals for a national repository were eventually dropped, the problem of dealing with nuclear waste had become a key issue in the conflict over the future of nuclear energy. In the UK and elsewhere, solutions to the problem have proved elusive both technically and politically.

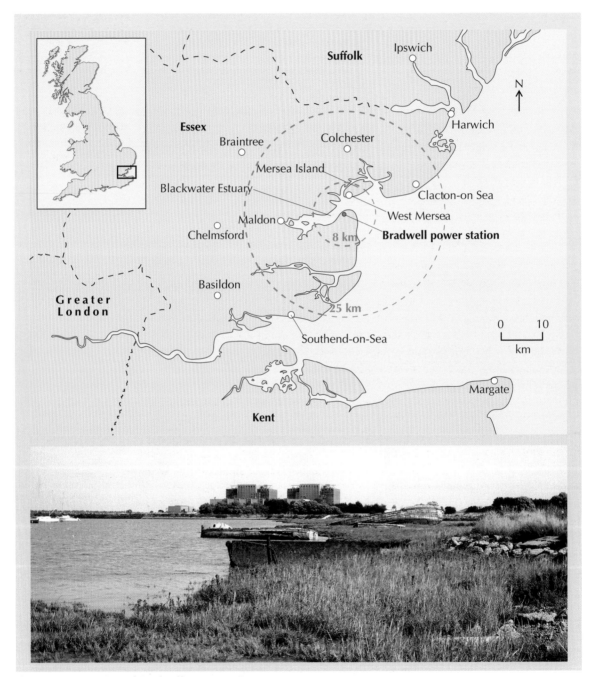

Figure 3.11 Location of Bradwell power station.

In this section we shall continue our focus on the concept of power, but this time in the context of the nuclear industry to examine how changing power relations affect those local communities where nuclear activities are concentrated. We shall keep in mind the two questions we began this chapter with: how and why do conflicts over the location of nuclear waste facilities arise, and what influence do changing power relations have on the outcomes of these conflicts?

3.2 Technical aspects of the problem

All nuclear power plants create radioactive wastes. These are dealt with in various ways according to their volumes, level of radioactivity and form. In the case of Bradwell, and most of the other operating nuclear plants in the UK, spent fuel from the reactor core is removed and transferred to Sellafield in Cumbria where it is reprocessed. Reprocessing is a form of chemical processing or recycling of spent fuel originally developed to recover uranium and plutonium for nuclear fuel for weapons production. However, reprocessing also produces wastes. For a representation of the scale of the problem, see Figure 3.13.

Figure 3.12 Protesting about the proposed nuclear waste landfill sites in the 1980s: LAND was Lincolnshire against Nuclear Dumping.

Although reprocessing does not change the amount of radioactivity present in nuclear materials, it does increase the volume and complexity of the resulting waste streams. In terms of radioactivity these wastes are categorized as low, intermediate and high (see Box 3.3). High Level Wastes (HLW) are cooled in ponds before being vitrified and stored: see Figure 3.14(a). Intermediate Level Wastes (ILW) are encapsulated and stored, while the bulky volumes of the UK's Low Level Wastes (LLW) are compacted and disposed of in a purpose-built repository at Drigg near Sellafield: see Figures 3.14(b) and (c).

Reprocessing has become controversial in recent years as the commercial and military reasons for producing plutonium have diminished. Commercially there are ample supplies of uranium ore, which is cheaper than reprocessed fuel, and, in an era of military decommissioning, there are ample stocks of weapons grade plutonium. As a result there is a problem of the potential proliferation of these materials which must consequently be safeguarded from seizure by terrorists or illegal diversion into the black market to prevent the potential development of nuclear weapons. In the case of Magnox fuel there are technical reasons for reprocessing since it corrodes in wet storage and becomes unstable, unsafe and expensive to manage without treatment.

The UK and France are the only western countries engaged in international commercial reprocessing operations (though Russia reprocesses fuel from eastern European reactors built in the former USSR). Apart from Magnox fuels which are dealt with separately, the UK's THORP (Thermal Oxide Reprocessing Plant) at Sellafield reprocesses fuels from the later AGR (Advanced Gas-cooled Reactor) stations and from PWRs (Pressurized Water Reactors) in other countries, notably Germany and Japan. By agreement, the reprocessed

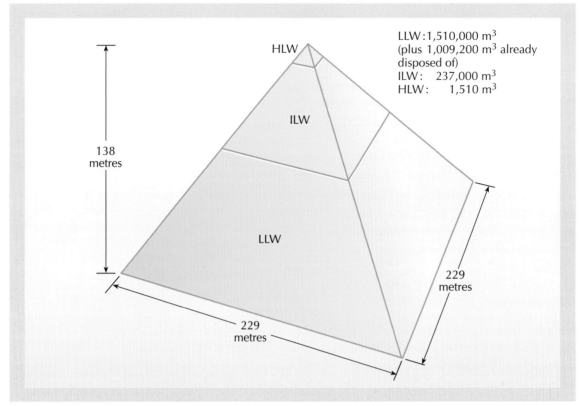

LLW:1,510,000 m³
(plus 1,009,200 m³ already
disposed of)
ILW: 237,000 m³
HLW: 1,510 m³

HLW

ILW

138
metres

LLW

229
metres

229
metres

Figure 3.13 Total volume of conditioned wastes, both existing and anticipated future waste, that will require long-term management after April 2001 (including waste already disposed of). To give a visual representation of the volumes of waste we have shown the Great Pyramid of Giza as this has a similar volume to the to total volume of waste (in fact 2,349,760 m³). The pyramid can be divided into different sections to represent the proportions of the volumes of the different categories of wastes.

Note: The volumes are quoted as conditioned waste volumes. i.e. the volume of waste assuming that it has been solidified (conditioned) into a passive safe form for long-term management. This is a useful concept as it enables comparisons to be made on a consistent basis. It should be noted, however, that the majority of wastes currently in stock in the UK have not at this stage been conditioned; most wastes continue to be stored in a raw or partly treated form.

Source: Data from Nirex, 2001 National Inventory (Nirex report N3/99/01).

plutonium and uranium will eventually be returned, perhaps in the form of fuel produced in the Sellafield MOX (Mixed Oxide fuel) plant together with the wastes arising. The alternative to reprocessing – now practised in most nuclear countries – is a once-through cycle whereby spent fuel is stored (either in ponds or dry stores). Whether in the form of spent fuel in stores or wastes arising from reprocessing, long-term solutions for the management of nuclear wastes will be necessary. Facilities for dealing with ILW and LLW will also be needed and volumes will increase as decommissioning of plant proceeds. Finding sites for solutions will be difficult, as the experience of Bradwell and communities in many countries has shown.

Box 3.3   Nuclear waste

Radioactive waste can be divided into three categories according to how radioactive it is – low, intermediate (medium) or high.

High-level waste has the greatest concentration of radioactive materials and produces substantial quantities of heat. This waste is what remains when the uranium and plutonium have been removed from spent (used) nuclear fuel rods through chemical separation (reprocessing). It is in liquid form and is cooled before being vitrified (converted into glass) and stored. In many countries spent fuel is not reprocessed and is therefore HLW. Surplus plutonium may also eventually be declared HLW.

Intermediate-level waste is far less radioactive than high-level waste. It is made up of such things as fuel element cladding, contaminated equipment and sludge that come from the treatment processes, although they can arise elsewhere when plant is dismantled. This solid waste can be encapsulated in cement inside steel drums and stored. Most of these wastes are stored at Sellafield.

Solid low-level waste, the least radioactive, includes paper towels, clothing and laboratory equipment from areas where radioactive materials are used. Decommissioning of nuclear power stations and other plant will produce large volumes of LLW in the form of building materials and redundant plant. Hospitals also produce low-level waste, as do some industries and research establishments. LLW in the UK is disposed of in containers inside a concrete vault at Drigg, near Sellafield and is also stored at Dounreay and other nuclear sites and facilities.

By way of comparison it is estimated that one tonne of used nuclear fuel produces the following volumes of waste after reprocessing:

0.1 cubic metres of HLW which contains 99% of the radioactivity;

1 cubic metre of ILW which contains 1% of the radioactivity; and

about 4 cubic metres of LLW which contains 0.001% of the radioactivity in the used fuel.

(a)

(b)

(c)

Figure 3.14 (a) Spent nuclear fuel in cooling pond, Sellafield (high level waste); (b) low level wastes stacked in containers in the disposal facility at Drigg, Sellafield; (c) moving the waste containers at Drigg.

Given the passions aroused in the 1980s **(Blowers and Smith, 2003)**, it is perhaps a little ironic that, since its closure in 2002, Bradwell has become a site where ILW arising from decommissioning the plant may be stored for at least a hundred years. Once a plant such as Bradwell has closed and its spent fuel has been removed to Sellafield, the process of decommissioning begins. This involves the demolition of those buildings, including the turbine hall, that are not radioactive and the recovery, packaging and storage on site of ILW (resins, cladding and other components of the reactor core). The reactor buildings will be clad in concrete 'safestore' structures while radioactivity levels decay. Just how long these structures will have to remain will depend on assessment of economic, technical (feasibility of options) and environmental (amenity) factors as well as risk to workers and the population at large, but it could be for a period of 80 to 100 years. This would mean that the final phase of decommissioning, the clearance and decontamination of the site and its restoration for other uses, will not be completed until well into the next century. The legacy of nuclear power will thus be visible at many sites long after its use and will require the safe management and monitoring of radioactive waste over extended timescales.

Activity 3.7

Why is radioactive waste a LULU?

The answer to this activity is given at the end of this chapter.

3.3 Political aspects of the problem

The hazards relating to radioactivity are discussed in detail later in this series (Hinchliffe and Blowers, 2003).

The process of radioactive decay results in the emission of different types of radiation which can cause immediate damage to living tissue if doses are high and either directly or indirectly increases the risk of hereditary defects and malignant disease. The statistical risk of a hazard occurring may be very small. However, any accident which does occur has potentially widespread and irreversible consequences. For these reasons, radioactive waste presents an example of a 'dread risk', the more so because it cannot be seen, felt or touched and is dangerous by mere proximity. It inspires great anxiety by virtue of the hazard to life and to the environment which it poses.

The fears generated by radioactive waste are sometimes dismissed by experts and those in the nuclear industry as unfounded and irrational. After all, they argue, the management of wastes in a future disposal facility should be designed to achieve a risk target to the public of 10^{-6}/year (i.e. one in a million) of developing either a fatal cancer or a hereditary defect (HMSO, 1995, para.78). Furthermore, the public fails to discriminate between the levels of risk presented by different nuclear activities. Yet it might equally be argued that these fears are both rational and realistic. Given the unimaginable timescales

Figure 3.15 A three-metre mutant fluorescent lobster emerging from a drum of radioactive waste was used by Friends of the Earth in one typically media-attractive campaign to draw attention to the stockpiling of nuclear waste at Sellafield from imports of spent nuclear fuel.

over which some nuclear wastes remain dangerous, the idea that precise probabilities of risk can be calculated seems itself irrational. It is difficult to predict institutional survival with any confidence over more than a few years (the frequent changes in the organization and ownership of the energy industry are evidence of this) and even geological predictions of beyond 10,000 years become increasingly uncertain.

The risks have been highlighted by environmental groups, notably Greenpeace and Friends of the Earth, through various actions such as surveying radioactivity on beaches or in shellfish, protesting against sea dumping and attempting to block the Sellafield emissions pipeline into the Irish Sea (see Figure 3.15). Governments, too, have become concerned and this has led to the international banning of dumping in the oceans **(Brandon and Smith, 2003)** and to a commitment to reduce discharges from land to sea: the undertaking, under the OSPAR (Oslo Paris) Convention, is that additional concentrations of radioactive substances in the marine environment above historic levels will, by the year 2020, be 'close to zero'.

The safe management of highly radioactive solid wastes has become a major environmental issue. The contest over the problem of risk posed by radioactive waste has generated political conflicts over their management that have, so far, prevented the development of long-term solutions. Political attention has become focused more on the end of the nuclear cycle (reprocessing and waste) than on the generation of nuclear energy (Figure 3.16). There seem to be four reasons for this. One is that these activities are perceived to create the highest risks, reprocessing being responsible for four-fifths of present exposures from the

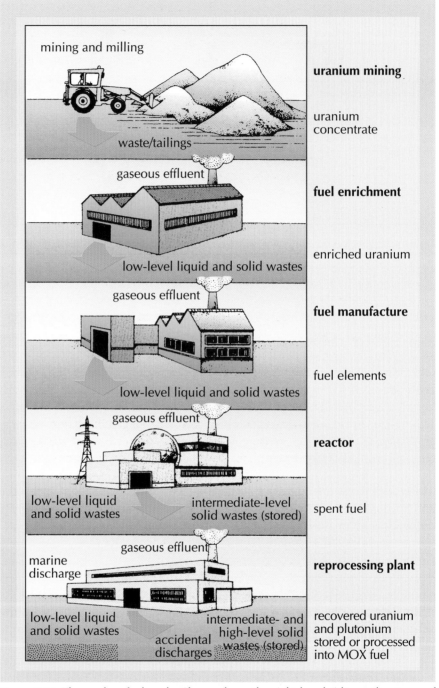

Figure 3.16 The nuclear fuel cycle. Shown down the right-hand side are the stages of the process and the output at each stage: this output forms the input into the next stage. The left-hand side highlights the discharges and wastes at each stage.

nuclear industry, and radioactive waste presenting risks extending down the generations. The second is that large volumes of waste have already accumulated and, as decommissioning proceeds, the scale of the problem will increase. The third is that with both military and civil nuclear production in decline, opposition to the nuclear industry has concentrated its energies on those areas where the industry appears to be most vulnerable. Anti-nuclear groups maintain the position that there can be no agreement on radioactive waste management solutions without the phasing out of the nuclear industry. And the fourth reason is that, in nearly every nuclear country, there is no long-term policy yet agreed for the management of nuclear wastes.

The future of nuclear energy is, in large measure, dependent on solutions being found to the problem of nuclear waste. But any solution must be located somewhere, and existing and potential host communities have become the focus of contests between interests with power resources changing over time.

3.4 Where is the problem?

From the analysis so far, what can we say about the location of radioactive wastes in the UK? Perhaps the most obvious conclusion is that, in terms of radioactivity, and to a lesser extent of volume, the wastes are remarkably concentrated spatially, at Sellafield (and the nearby Drigg LLW facility) and, to a much lesser extent, at Dounreay in Scotland (where a variety of wastes are stored resulting from a range of activities which have now closed): see Figures 3.17 and 3.18. This pattern will be modified but not substantially changed as plants such as Bradwell are decommissioned, leaving some ILW to be stored on site. Moreover, apart from the LLW repository at Drigg, the wastes will continue to be stored, monitored and kept under strict surveillance on site until a permanent method for their management is found. There is, therefore, a twofold problem: how to ensure the safe storage of accumulating wastes (and the plutonium arising from reprocessing) and how to find a long-term policy for their management.

Radioactive wastes and reprocessing facilities tend to be found in what have been termed 'peripheral' communities (Blowers and Leroy, 1994). As we have seen,

Figure 3.17 The Sellafield plant in Cumbria, England.

Figure 3.18 The Dounreay plant in Caithness, Scotland.

remoteness was a characteristic of early siting decisions and the pattern has persisted whereby major nuclear facilities (though not all power plants) are distant from major populations or are relatively inaccessible. Examples of such peripheral communities are found in places – like Sellafield in Cumbria – where a variety of nuclear operations provide a large employment base in relatively remote areas. In the United States, Hanford, on the Columbia River in the semi-desert of Washington state, is another example: see Figure 3.19. It began its life during the Second World War, engaged in highly secret plutonium production for the development of the atomic bomb. It later expanded with a range of experimental facilities and reactors, to become by far the largest employer in the Tri-Cities (made up of the cities of Richland, Kennewick and Pasco) area with 100,000 population. Hanford's huge site, covering 1658 square kilometres, is notoriously contaminated as a result of early experimental and accidental releases and the problem of leaking wastes from liquid HLW stored on the site.

The reprocessing complex at Cap de la Hague, on the tip of the Cotentin peninsula near Cherbourg in France (see Figure 3.20), is the dominant employer in a remote corner of Normandy and also exhibits the remoteness characteristic of peripherality:

> Picture a harsh, wild stretch of nature with splendid horizons, endlessly fascinating in terms of the variety it offers to the eye. The interior of the peninsula is occupied by a great plateau consisting of dome-like moors where gorse and broom, heather and bracken are swept by incessant wind

(Zonabend, 1993, p.13)

Peripherality is not simply a spatial concept; it has economic, political, social and environmental connotations, too.

Economically, nuclear plants are often the dominant employer, resulting in the high dependency of the local population on the investment and high wages generated by the industry. Politically, this condition of dependence can sometimes lead to a sense of powerlessness. Local political élites must have regard to local prosperity and the need to protect employment, so in West Cumbria the local council, Copeland, has generally adopted a supportive position to the nuclear industry. Cumbria County Council, however, represents a much wider geographical area and therefore a wider variety of interests, and so has tended to promote environmental considerations. The local sense of powerlessness stems from the fact that key decisions affecting the area are taken

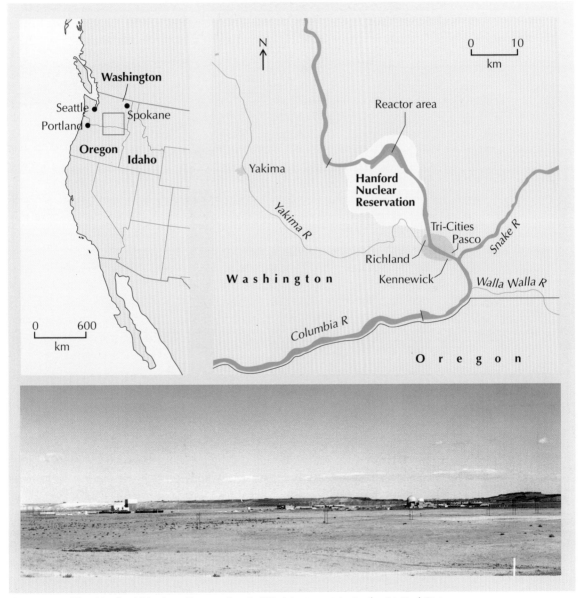

Figure 3.19 The Hanford Nuclear Reservation in Washington state in the United States.

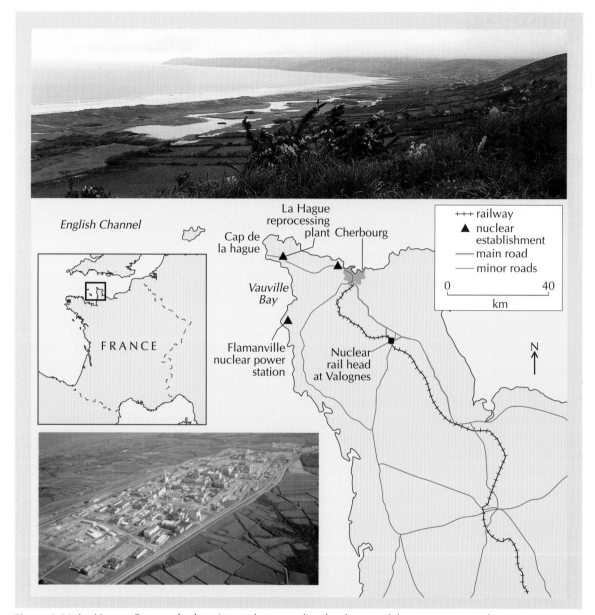

Figure 3.20 La Hague, France: the location and surrounding landscape of the reprocessing plant.

elsewhere by national government and the nuclear industry. The Sellafield plant is run by British Nuclear Fuels (BNFL) whose headquarters are at Risley near Warrington and who operate 51 sites in sixteen countries around the world.

In social terms, communities such as Sellafield display a culture of defensiveness and an ambiguity towards the industry. A study undertaken by Lancaster University depicted the condition at Sellafield as follows: 'People combined a sense of fatalism about the situation with indications of guilt at being a "community" which allowed itself to be dictated to so comprehensively, not just by the nuclear industry but by outside forces more generally' (Wynne et al., 1993,

Figure 3.21   Security at a nuclear site.

p.37). The feelings of separation and isolation can be exacerbated by the fact that nuclear facilities are heavily protected to prevent unauthorized intrusions and, in some cases, to seal off contaminated areas (Figure 3.21).

Finally, these peripheral communities are usually associated with some form of environmental degradation as a result of hosting the nuclear facility. The nuclear industry can create visual intrusions (such as Sellafield on the edge of the Lake District, or Bradwell imposing its bulk on the Blackwater) which damage

Figure 3.22   Three containers for transporting nuclear waste for underground storage to the Waste Isolation Pilot Plant near Carlsbad, New Mexico, pictured in the background.

amenity, at least for some people. Yet it is the potential contamination of habitats and the impacts of radioactivity on the health of species, including humans, that constitute the most significant problems. For example, the leaking of high level liquid wastes from tanks on the Hanford site poses a potential hazard to the Columbia river.

3.5 Power and the periphery

Hanford (begun in 1943) and la Hague (1966), along with Sellafield (1947), are well-established sites. Each community, in different ways and to varying degrees, exhibits the beleaguered nature of peripherality, a condition of powerlessness in a context where their fate is decided elsewhere. Occasionally new sites can be established. For example, a deep disposal facility has been opened near Carlsbad in the semi-desert of New Mexico, USA, in an area where the collapse of the phosphate industry had left the community economically vulnerable (see Figure 3.23). But powerlessness is not an inevitable condition even in established sites, as we shall see. It is much more difficult to find new sites where the conditions of peripherality are not met. We have noted how, in the 1980s, the attempts to establish greenfield sites for nuclear waste facilities in eastern England were thwarted **(Blowers and Smith, 2003)**. Elsewhere, in Gorleben, in Germany, efforts to establish, first, a reprocessing complex and, later, a repository for high level wastes, have met with intense resistance.

Activity 3.8

What are the five characteristics of peripheral communities?

The answer to this activity is given at the end of this chapter.

The conflicts over the location of nuclear waste provide us with some insights into power and powerlessness. There are various interests at stake in these conflicts and each has access to specific power resources. The nuclear industry possesses power through the wealth creation that provides employment and local investment. This power is sustained by virtue of its dominance of the local economy and the fear (sometimes made palpable by the threat) of withdrawal. In this respect the industry's access to government helps to ensure that its interests are met. Thus, the power of the industry exists, in a sense, precisely because of its presence – the local community has nowhere else to turn.

The power of the industry is by no means uncontested, however, even in its most established locations. Despite its powerful presence the industry is also a focus of opposition as a LULU which brings risk and environmental degradation to a locality. This weakness becomes most exposed when the industry seeks new locations in places where it lacks an existing power base. In these situations the industry encounters powerful local resistance from communities anxious to protect their environment. As in the case of eastern England and Gorleben, these

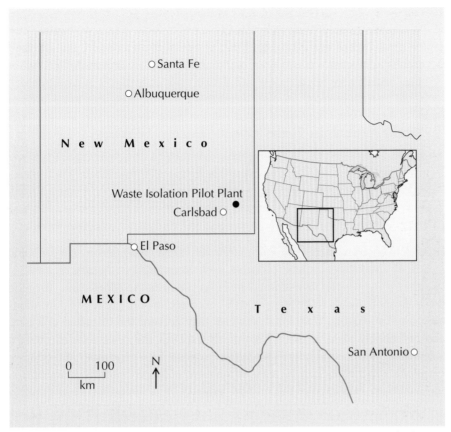

Figure 3.23 The location of the Waste Isolation Pilot Plant (WIPP) near Carlsbad, New Mexico, in the United States.

communities may win the support of broader environmental movements able to gain the attention of local and national media for their campaigns. Environmental groups have strengthened their power resources through attracting public support and gaining access to policy-makers.

These examples suggest that the relative power of the opposing interests is dynamic, changing over time in response to economic and environmental circumstances. The power of interests relates to the discourse of power prevailing at a particular time. In the early years, when there was optimism about the nuclear future, the industry was able to establish itself in remote locations relatively unopposed. Later, as fears grew about environmental risks – fanned by a number of incidents such as Windscale (now Sellafield) in 1957, Three Mile Island, USA, in 1979 and, most of all, by Chernobyl in the Ukraine in 1986 – the industry was placed on the defensive. Its vulnerability increased as its commercial position waned and the industry declined in most countries. Over time, the relative power of the nuclear industry has diminished and it has experienced great difficulty in establishing new locations. Therefore if there is further nuclear development, it may be expected either at existing nuclear sites or in economically marginal locations.

This analysis suggests that peripheral communities are not as powerless as they might at first appear. Taking each aspect of peripherality in turn we can see how power relationships are changing. First, in terms of remoteness, though their geographical position cannot alter, communities can become more accessible. Physical access may remain slow and difficult (for example, it is hard to reach Sellafield by train or road); in some cases there are long distances to the nearest major communities (Hanford is about 240 km from Spokane and 400 km from Seattle and Portland: see Figure 3.19); the area may be isolated at the end of a peninsula (la Hague) or on the borders of a country (as was Gorleben before German unification: see Figure 3.24). But, while physical remoteness undoubtedly presents problems, electronic means of communication increasingly ensure integration into the wider community, enabling these communities to draw on powerful resources elsewhere.

Second, in terms of their economic marginality, their dependence on a declining source of employment tends to enable such communities to claim recognition as places in which investment, both in nuclear activities and through diversification, needs to be protected. By contrast, in some greenfield locations the power of existing traditional activities may prevent the intrusion of a new, and perceptibly alien, activity. This partly explains the resistance exerted by farming and other local interests against the plans for a repository at Gorleben. Economic marginality can be seen as a source of power whether in support of the continued existence of the nuclear industry or in opposition to its possible development.

Third, these communities possess political resources. On the one hand, the nuclear industry can count on the support of its workforce and continues to exert considerable influence with governments both locally and nationally. Indeed, the industry may not need to exert its power in any overt action: its mere presence and reputation may be sufficient to ensure its interests are heeded. However it is deployed, the source of the industry's power is economic. On the other hand, opponents of the industry are often able to build coalitions between different groups in the community, whether traditional such as farmers (Gorleben) or new incoming groups (Sellafield) and with other communities or environmental groups elsewhere sometimes acting at a national level. Their source of power is essentially political, deriving from the concern felt about the environmental risks of nuclear activities and the consequent ability to exert pressure on decision-makers through the media, lobbying and protests.

Fourth, these peripheral communities reflect the changing discourse of the nuclear industry. Culturally, they have moved from being part of a new, exciting and powerful technology through a period of defensiveness as they bore the stigma of an unpopular industry. But, with the industry having fewer locational options, the peripheral communities have acquired a greater significance (and potentially confidence) as they recognize their role in hosting activities unwanted elsewhere. In social terms the conflicts of interests surrounding these communities reveal the ability of both sides to assemble powerful coalitions

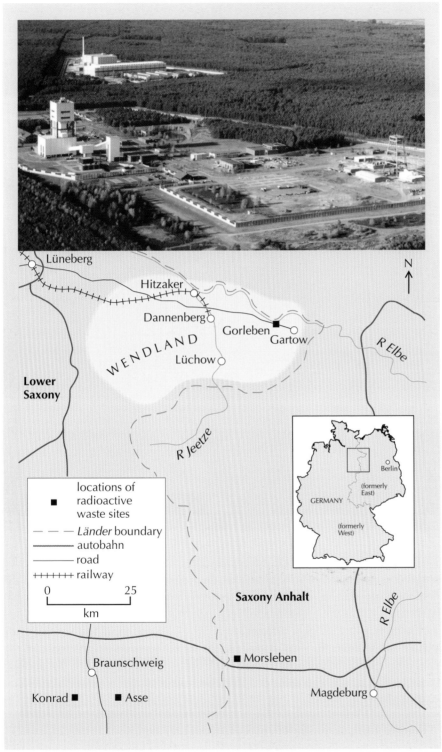

Figure 3.24 The interim nuclear waste store and potential repository site at Gorleben in Germany.

that cut across conventional social and geographical divisions. The alliance between capital and labour on which the strength of the industry rests is, at times of conflict, confronted by coalitions combining different classes, political allegiances, interest groups and communities. For example, at Gorleben, traditional and conservative farming and local interests linked up with radical protesters from other parts of Germany creating blockades of tractors, disruption of rail movements, peaceful demonstrations and sporadic clashes with police. The construction and eventual commissioning of an interim store for HLW was only achieved by massive deployments of armed police (Figure 3.25).

Figure 3.25 Protesters at Gorleben, Germany.

Last, the environmental risk that is associated with the nuclear industry in peripheral communities also influences their power relationships. Whereas the benefits from nuclear energy are dispersed throughout society, many of the hazards in terms of contamination are relatively localized. Communities are increasingly aware of the environmental risks they are being asked to bear on behalf of society. But as this has become increasingly recognized by environmental groups, government and industry, so the need to address these environmental problems has itself become a source of power to the local communities.

Activity 3.9

In what ways have power relations in nuclear communities been influenced by changes in (a) the power resources available to different interests and, (b) the changing discourse of power concerning nuclear energy?

Comment

In general terms it may be said that, over time, the power resources of the nuclear interests have weakened as the industry has declined. However, the economic vulnerability of the workforce has, in places such as Sellafield and Hanford, ensured continuing investment in the industry in its main locations. Anti-nuclear interests have experienced increasing power resources through their access to the media and government and their ability to mobilize coalitions of protest linking local and broader interests. These changing power relations also reflect the changing discourses of nuclear energy. The optimism which characterized the early years has been succeeded by a discourse of risk as a series of accidents, resulting in public recognition of the risks, created a lack of confidence and trust in the industry.

In most nuclear countries strategies are being developed which have to take account of the changing power relations affecting the nuclear industry. It has been recognized by most countries that the management of radioactive waste at specific locations cannot be based on exploiting disadvantaged communities; rather the rights of communities must be acknowledged both to participate in decision-making and to be compensated for accepting the risks created by society as a whole. Some countries are moving towards a system whereby volunteer host communities are sought for the location of radioactive wastes. The success of the volunteer principle will depend on the package of incentives, compensation and participation that is offered. Among the possible incentives are tax benefits, financial investment and employment guarantees; compensation embraces such matters as infrastructure provision, community facilities and environmental enhancements; while participation includes involvement in decision-making with the possible right of veto over decisions. In some cases, the compensation may be rejected as inadequate or regarded as a bribe; sometimes the price demanded may prove too high. For example, the community of Deep River in Canada (adjacent to the Chalk River research and reactor complex) was willing to host a repository for LLW provided that employment would be sustained at 1995 levels, a condition that the federal government was unwilling to meet.

Using the volunteer principle combined with compensation measures, progress has been made towards finding an acceptable site in Finland and the volunteer approach has been adopted by Sweden, Switzerland, Spain and France. Success or failure of such an approach will depend on a range of circumstances affecting individual countries and sites. But the approach indicates that social and political criteria are as important as scientific and technical criteria in seeking solutions to the problem of radioactive wastes.

3.6 Reflections on the nuclear case study

In this case study we have focused on power relations in the context of peripheral communities to answer the two questions we began this chapter with: how and why do conflicts over the location of nuclear waste facilities arise, and what influence do changing power relations have on the outcomes of these conflicts? These communities have been dependent on the employment and wealth created by the nuclear industry which met little opposition in its early years. But, as the industry has declined, the focus has turned towards the problems of environmental risk arising from reprocessing and radioactive wastes. Conflicts have developed at existing sites and at proposed locations revealing a changing pattern of power relations. In terms of the key analytical concepts we are using, the nuclear case study reveals some interesting and potentially important aspects of the relationship between places, power and action. Although nuclear power generation is a high-tech modern industry, it has characteristics of scale and technology that are typical of more traditional industrial processes such as steel and heavy engineering. In a social sense, too, the conflicts over the nuclear industry, and latterly over nuclear waste, invigorate traditional forms of community cohesion and integration. Within these peripheral nuclear communities, industry and workers, capital and labour, can be obdurate in their defence of investment and jobs, as has been the case at Sellafield.

Over time, power relations within nuclear communities have been changing. It is true that such communities remain economically vulnerable and areas of environmental risk. But this very vulnerability provides a source of cohesion and power. Since withdrawal of investment would have a devastating impact on the local economy, the isolation of these communities ensures their continuing survival. In any case, in places such as Hanford, Dounreay and eventually Sellafield and la Hague, decommissioning and clean-up maintains a substantial workforce. But, as the locational options for the future management of nuclear wastes have narrowed, so the industry has to turn to its established bases. So, in answer to our second question, we can see that changing power relations hold the key to understanding the apparent locational inertia of the nuclear industry. The community's dependence on the industry is increasingly matched by the industry's dependence on the community, a union of mutual dependency. In return for their continuing management of society's problem, these communities are able increasingly to exert pressure for compensation and a greater say in their future.

In this study we have concentrated on the spatial dimension and have said very little about the temporal dimension. Yet radioactive waste, more than anything else, confronts us with dealing with a problem whose potential consequences extend into the far future. This feature raises important issues concerning risk, uncertainty, sustainability and democracy, issues which we shall be considering in the final book in this series **(Blowers and Hinchliffe, 2003)**.

Activity 3.10

Thinking back over the chapter, can you summarize the similarities and contrasts between wind and nuclear energy in terms of their relationships to local communities?

Comment

Similarities include the tendency for both wind farms and nuclear facilities to be located in relatively remote areas. In the case of wind farms this is because windy upland or coastal sites are the most appropriate locations. Nuclear facilities are located in remote areas because of the risks and political objections they raise. Both wind farms and nuclear plants are regarded as LULUs and attract opposition from communities because of the visual intrusion and blight that they cause.

However, there are considerable differences. The main problem with wind farms is amenity damage (appearance and noise) whereas, in addition to this, nuclear facilities present long-term risks through radioactivity to both the environment and species. Wind farms create very little local employment whereas the nuclear industry is often the dominant employer in a community. In terms of location, wind is a developing and distributed form of energy with an increasing number of dispersed locations on land and offshore. By contrast, nuclear energy, though it remains a significant source of electricity and employment, is confined to well-established locations. Opposition to wind farms tends to be limited to the localities where they are proposed and usually wanes once they are established; opposition to nuclear facilities, ventilated by environmental groups, is more widespread and persistent.

4   Conclusion

In seeking answers to the questions of how and why conflicts over the location of wind farms and nuclear waste projects arise, the two cases presented here offer some interesting contrasts and parallels. Both wind farms and radioactive waste projects arouse considerable opposition from prospective host communities. We have, for instance, indicated the degree of opposition expressed against wind farms in Wales and proposals for nuclear waste repositories have been vigorously resisted in, for example, eastern England, at Gorleben in Germany as well as at Sellafield. There are some similarities to the grounds for objection. Wind farms and nuclear waste projects impose environmental costs on the local community without corresponding benefits: they are, in short, LULUs. Moreover, these communities experience a sense of resentment at their powerlessness in the face of external forces largely beyond their control. Some of the wind protesters see themselves as fighting against powerful vested interests, including ultimately the government, and therefore share something in common with the more conventional environmental campaigners, including anti-nuclear campaigners. So, to some extent, they may share a sense of powerlessness as well as sometimes adopting similar tactics in response.

Yet there are some important differences. At the national level the majority of people asked in the UK generally *support* the use of wind energy (between 64 and 91 per cent in Table 3.1), while a similar percentage typically *oppose* or are concerned about nuclear power. As Figure 3.26 shows, the latter has varied with time and, as one might expect, increased immediately after the Chernobyl accident in 1986. As can be seen, it also increased in 1989/90 when it became apparent, as a result of the attempt to privatize the nuclear plants, that nuclear power was more expensive than had previously been suggested.

More recently, in a poll carried out by the British Market Research Bureau in 2001, 68 per cent of those interviewed said that they 'did not think that nuclear power stations should be built in Britain in the next ten years' (BMRB, 2001). A year later, a MORI poll of the UK residents found that 72 per cent of those asked favoured renewable energy rather than nuclear, while a National Opinion Poll carried out in the same year for the Energy Saving Trust found that 76 per cent believed the government should invest time and money developing new ways to reduce energy consumption, 85 per cent wanted government investment in 'eco-friendly' renewable energy (solar, wind and water power) and only 10 per cent said the government should invest time and money in building new nuclear plants.

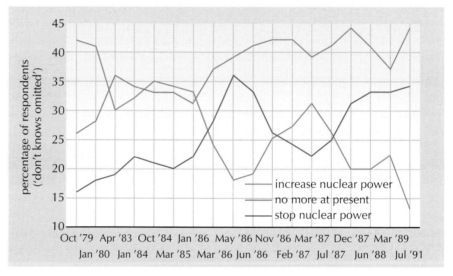

Figure 3.26 Public opinion on nuclear power in the UK, 1979–91.
Source: Social Surveys (Gallup) from the UK Government's Nuclear Review; submission by Stop Hinkley Expansion, September 1994.

These differences go some way towards explaining the outcomes of conflicts over wind farms and nuclear waste projects. Although wider concerns are sometimes apparent, most of the local wind farm protests in the UK seem to be NIMBY in nature. They are mainly protests over local amenity and, once wind farms have been developed, the bases for conflict tend to diminish as the projects gain acceptance. Indeed, the widespread support for wind energy precludes continuing conflict. By contrast, protests over nuclear waste facilities focus on the environmental and health risks, both local and further afield, and extending into the far future. These concerns have brought together both local communities and a much wider constituency of anti-nuclear interests in powerful coalitions able to use local and national media support and political leverage to defeat proposals. In contrast to wind farms, nuclear waste projects have, so far, failed to gain broad public acceptability.

In terms of action, in both cases local campaigners have been able to enlist the support of the local media – who always seem to be keen to cover local conflicts. In addition both have, on occasions, made use of more direct methods of opposition – although few wind farm opponents have resorted to the more militant direct action tactics adopted by some anti-nuclear campaigners operating on a national scale as, for example, at Gorleben. The wind protesters have usually limited themselves to posters, letters, petitions and lobbying, rather than to major demonstrations and blockades.

The interests supporting wind energy projects can draw on the widespread public support revealed in the opinion polls. This support, together with government backing, ensures that wind projects materialize, although local opposition has impeded progress and diverted wind farms to other locations including offshore. By contrast, the nuclear industry has to rely heavily on the local support it can find in its traditional bases or in areas of economic decline such as Carlsbad, New Mexico. In places such as Sellafield, la Hague and Hanford the interdependence of community and industry, investment and jobs makes continuing economic prosperity a powerful counter to environmental concerns. Although this union of mutual dependency may ensure the survival of existing activities, the development of new nuclear waste facilities, which bring few jobs but many long-term risks, is likely to be strongly contested.

The relative power available to the protagonists may be explained in terms of our second question: what influence do changing power relations have on the outcome of these conflicts? From a position of strength in the early days the nuclear industry has been confronted with an increasingly confident and powerful opposition able to use and develop a discourse of environmental risk and fear. This has diminished the prospects of finding acceptable solutions to the problem of nuclear waste. Wind energy has developed a much more positive discourse as an environmentally-friendly renewable source of energy. While this enhances the prospects for development overall, local opposition can still, in some cases, prevent projects in some locations.

In both cases solutions will need to focus on local concerns. While it is conceivable that the advent of wind co-ops and locally owned wind projects of the type that exists widely in Denmark and Germany could help reduce opposition to wind projects in the UK, it is hard to imagine how there could be local ownership of nuclear waste repositories, although, as we have seen, local communities near nuclear projects are to some extent bound to them by employment and other economic benefits. Thus compensation for local disbenefits and positive local economic and employment benefits from the projects can be relevant in both cases as ways in which to reduce local opposition and to win approval. This is an issue we shall take up in the next book in the series **(Blowers and Hinchliffe, 2003)**.

Meanwhile, we leave you to ponder this question: what would be *your* response if either a wind farm or a nuclear waste facility was proposed near you? Your answer may help you to understand how and why conflicts arise and how power relations influence outcomes.

References

Allen, G. (1991) *The Scramble for Windfarms*, Northern Devon Group of the Council for the Protection of Rural England.

Aubrey, C. (1993) *The Guardian*, 5 November.

Blowers, A. and Leroy, P. (1994) 'Power, politics and environmental inequality: a theoretical and empirical analysis of the process of "peripheralisation"', *Environmental Politics*, vol.3, no.2, Summer, pp.197–228.

Blowers. A.T. and Hinchliffe, S.J. (eds) (2003) *Environmental Responses*, Chichester, John Wiley & Sons/The Open University (Book 4 in this series).

Blowers, A.T. and Smith, S.J. (2003) 'Introducing environmental issues: the environment of an estuary' in Hinchliffe, S.J. et al. (eds).

BMRB (2001) Public opinion survey carried out for the Royal Society for the Protection of Birds, British Market Research Bureau.

Brandon, M.A. and Smith, S.G. (2003) 'Water' in Morris, D. et al. (eds).

CCW (Countryside Council for Wales) (1992) *Wind Turbine Power Stations*, Cardiff, Countryside Council for Wales.

CPRE (Council for the Protection of Rural England) (1991) Evidence to the Department of Energy's Renewable Energy Advisory Group.

DEFRA (2000) *Guidance on Preparing Regional Sustainable Development Frameworks*, February, Department for Environment, Food and Rural Affairs, http://www.defra.gov.uk/environment/sustainable/rsdf/guidance2000/index.htm [accessed February 2003]. (Note that this guidance is now categorised as a DEFRA paper although DEFRA did not exist in 2000.)

DETR (2000) *Climate Change: UK Programme*, Department of the Environment, Transport and the Regions, London, HMSO.

ETSU (1993) *Attitudes towards Windpower: A Survey of Opinion in Cornwall and Devon*, Energy Technology Support Unit, W/13/00 354/038/REP.

Gipe, P. (1995) *Wind Energy Comes of Age*, Chichester, John Wiley & Sons.

Harper, M. (1994) 'Wind energy still blowing strong', *Safe Energy*, no.99, February/March, Edinburgh, SCRAM.

Hinchliffe, S.J. and Belshaw, C.D. (2003) 'Who cares? Values, power and action in environmental contests' in Hinchliffe, S.J. et al. (eds).

Hinchliffe, S.J. and Blowers, A.T. (2003) 'Environmental responses: radioactive wastes and uncertainty' in Blowers, A.T. and Hinchliffe, S.J. (eds).

Hinchliffe, S.J., Blowers, A.T. and Freeland, J.R. (eds) (2003) *Understanding Environmental Issues*, Chichester, John Wiley & Sons/The Open University (Book 1 in this series).

HMSO (1955) *A Programme of Nuclear Power,* Cmnd 9389, London, HMSO.

HMSO (1995) *Review of Radioactive Waste Management Policy*, Cm 2919, London, HMSO.

Morris, D., Freeland, J.R., Hinchliffe, S.G. and Smith, S.R. (eds) (2003) *Changing Environments*, Chichester, John Wiley & Sons/The Open University (Book 2 in this series).

NATTA (1993) *Renew*, no.85, September–October, Milton Keynes, Network for Alternative Technology and Technology Assessment.

NATTA (1994) 'Story of ugliness of wind turbines: wind turbines – the truth', *Factsheet* (dated 1993) reprinted in *Renew*, no.87, January–February, Milton Keynes, NATTA.

NATTA (2001) 'Public attitudes towards wind farms in Scotland', System Three Social Research (as reported in *Renew*, no.130, Milton Keynes, NATTA).

Openshaw, S. (1986) *Nuclear Power: Siting and Safety*, London, Routledge and Kegan Paul.

Reddish, A. (2003) 'Dynamic Earth: human impacts' in Morris, D. et al. (eds).

Shell International (1995) *The Evolution of the World's Energy System 1860–2060*, London, Shell International.

Toke, D. (1999) 'Community ownership – the only way forward for UK wind power?', NATTA Report, Milton Keynes, NATTA.

Walker, S. (1993) 'Down on the windfarm', NATTA Report, Milton Keynes, NATTA.

Welsh Affairs Committee (1994) House of Commons Welsh Affairs Committee Session 1993–94, Second Report, *Wind Energy*, vol.1, London, HMSO.

Wynne, B., Waterton, C. and Grove-White, R. (1993) *Public Perceptions and the Nuclear Industry in Cumbria,* Lancaster, Centre for the Study of Environmental Changer.

Zonabend, F. (1993) *The Nuclear Peninsula*, Cambridge, Cambridge University Press.

Answers to Activities

Activity 3.1

You should have noticed that there are two different measures being used here: the 6.6% figure is the percentage of the world's total *primary energy*, supplied by nuclear plants, while the 25% figure is the percentage of the UK's *electricity* that is produced by nuclear plants. Remember that electricity, which is what nuclear power plants produce, is only one of the forms of energy that people use. Nuclear power supplies around 17% of the world's *electricity*, so the UK does generate more than the world average.

Activity 3.2

The main disadvantage is the variability of the source of supply: it is not always sunny or windy. While biomass can be stored (it is, in effect, stored solar energy), most of the other renewable energy sources have the disadvantage that natural energy flows tend to be dispersed, diffuse and often intermittent – they rely on the climate and weather systems, or predictable, but cyclic, phenomena like the tides and may be locationally specific. That may make it hard to supply power reliably on a larger scale, where and when we need it.

However, Denmark gets 18 per cent of its electricity from wind, and it has had no problems with the intermittency of this energy source: the national grid system averages out and buffers the effects of local variations in wind availability. For larger wind contributions, at some point in the future, some form of energy storage back-up would be required, possible involving a switch over to the production and use of hydrogen as a new fuel.

Activity 3.4

Wind turbines impose problems of noise and visual intrusion on local communities and these may impact on house prices, while the benefits tend be felt elsewhere, causing resentment that profits are going to big business.

Activity 3.5

Such compensation might set a precedent thus opening up the possibility that other groups would also seek compensation for living close to LULUs. An example of this might be those communities living near nuclear waste facilities, an issue discussed later in the chapter. It would also raise questions of who to compensate and the extent of the area affected.

Activity 3.7

They are locally unwanted land uses because the radioactivity present in the wastes presents a potential hazard to those communities living in close proximity to existing or proposed storage or disposal facilities.

Activity 3.8

The peripheral communities where nuclear facilities are located are:

> geographically remote
> economically marginal
> politically powerless
> culturally defensive, and
> environmentally degraded.

Chapter Four

Troubled waters

Pam Furniss

Contents

1   Introduction

'Of all the social and natural crises we humans face, the water crisis is the one that lies at the heart of our survival and that of our planet Earth.'

(Koïchiro Matsura, Director General, UNESCO, 2003)

The purpose of this chapter is to examine this 'water crisis' – what it is and the reasons for it – in the context of contest and conflict, the theme of this book. In the terms of the book's key questions, the main focus will be: how and why do contests over environmental issues – in this case, water – arise? Several contributory answers to this complex question will be considered.

The importance of water resources cannot easily be overstated, to us as individuals and to communities from local through regional and national to international scales. Various examples and case studies are included in this chapter to illustrate how contests over water can arise at these different levels. Some of the examples also demonstrate how conflict can be avoided. In this way the second key question, about environmental understandings, is addressed. It is interpreted here as: how can potential conflict over water be avoided?

The third key question asks: can environmental inequalities be reduced? Unequal access to water and to power resources is a recurring issue in this chapter, but it will be seen that it is not always easy to suggest opportunities to reduce these inequalities.

This chapter focuses mainly on the first two questions and develops them through each section. Sections 2 and 3 discuss several different factors that come into play as possible causes of water conflict and suggest practices and agreements that can improve understanding. Section 4 moves to the international level and describes real and potential water contests between countries.

2   Less for more

The main cause of the water crisis that faces us is simply that there is less available water in the world for more and more people. Before examining that in a little more detail, I want to consider the key concept of *value*, as it relates to water. In order to understand the reasons for contests over water, we need to recognize that it can be valued in many different ways.

2.1 What is water worth to you?

According to my own family's water bill, the cost of water for our household is less than £1 a day. For that, we (two adults and two teenage children) get an unlimited supply of clean, safely drinkable water that we can use for all our domestic purposes including cooking, washing and watering the garden, as well as drinking. It also pays for our waste water to be removed and treated. It seems like a very good deal to me. But be careful – £1 a day is what it costs me, it does not describe its value to me. If I paused to think about what my life would be like without an instantly available, clean water supply then I would place a very high value on it and probably be willing to pay more money for it too. We all depend on water to survive and yet most of us rarely think about its real value.

Figure 4.1 shows some comments I received from various sources in response to my question: what is water worth to you?

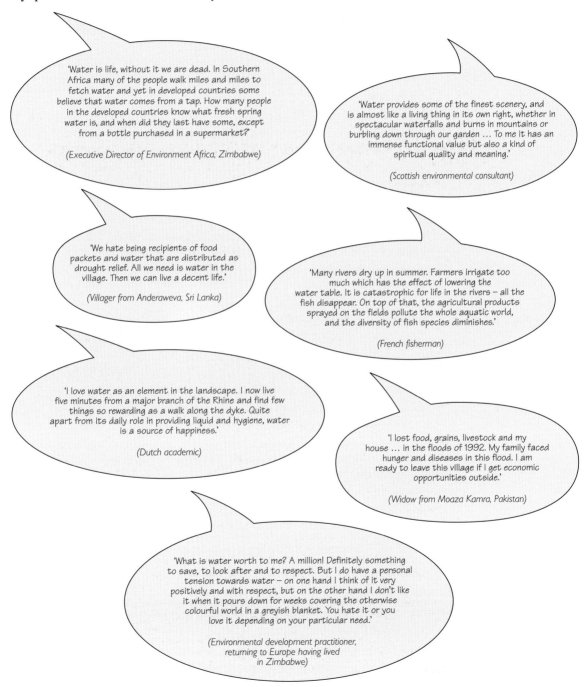

'Water is life, without it we are dead. In Southern Africa many of the people walk miles and miles to fetch water and yet in developed countries some believe that water comes from a tap. How many people in the developed countries know what fresh spring water is, and when did they last have some, except from a bottle purchased in a supermarket?'

(Executive Director of Environment Africa, Zimbabwe)

'Water provides some of the finest scenery, and is almost like a living thing in its own right, whether in spectacular waterfalls and burns in mountains or burbling down through our garden ... To me it has an immense functional value but also a kind of spiritual quality and meaning.'

(Scottish environmental consultant)

'We hate being recipients of food packets and water that are distributed as drought relief. All we need is water in the village. Then we can live a decent life.'

(Villager from Anderaweva, Sri Lanka)

'Many rivers dry up in summer. Farmers irrigate too much which has the effect of lowering the water table. It is catastrophic for life in the rivers – all the fish disappear. On top of that, the agricultural products sprayed on the fields pollute the whole aquatic world, and the diversity of fish species diminishes.'

(French fisherman)

'I love water as an element in the landscape. I now live five minutes from a major branch of the Rhine and find few things so rewarding as a walk along the dyke. Quite apart from its daily role in providing liquid and hygiene, water is a source of happiness.'

(Dutch academic)

'I lost food, grains, livestock and my house ... in the floods of 1992. My family faced hunger and diseases in this flood. I am ready to leave this village if I get economic opportunities outside.'

(Widow from Moaza Kamra, Pakistan)

'What is water worth to me? A million! Definitely something to save, to look after and to respect. But I do have a personal tension towards water – on one hand I think of it very positively and with respect, but on the other hand I don't like it when it pours down for weeks covering the otherwise colourful world in a greyish blanket. You hate it or you love it depending on your particular need.'

(Environmental development practitioner, returning to Europe having lived in Zimbabwe)

Figure 4.1 'What is water worth to you?'

Activity 4.1

Look at the comments in Figure 4.1. What do they reveal about the value these people place on water? Can you identify any groupings among these people?

Comment

Almost all of these quotes seem to place a high value on water in terms of its importance in their lives but they focus on different aspects. Some people indicate that the top priority and overriding concern is simply to obtain adequate quantities of clean, safe water in order to survive. For these people, the instrumental value of water as an essential resource is all important. There are others for whom access to water is apparently not a problem. It is noticeable that this second group all live in European countries and, while they acknowledge the fundamental necessity for water, they consider its aesthetic value is also important. It is possible that valuing water for the personal pleasure it gives you might also be true for the first group but, as far as these quotes reveal, this group have higher priorities that dominate over other, less pressing, needs. Another view is offered by the French fisherman who is aware of the ecological value of rivers for wildlife, especially the fish. He also raises the issue of conflict between different users because he blames farmers for damaging river life. The one exception to these generally positive views of the value of water is that of the widow from Pakistan for whom the destructive power of flooding reveals another side of human interaction with water.

We can see, then, that the value we give to water and its importance in our daily lives are going to depend on who we are, where we live and what we do. Recognizing these differences and appreciating that water plays many different roles in our lives begins to explore the first key question for this chapter: how and why do contests over water arise? With such a diversity of attitudes and priorities, coupled with the fact that water is essential for life, it is hardly surprising that the availability and use of water can be a contentious issue.

Attitudes and values have changed over time as well. In 1908, six years after the completion of the Low Aswan Dam on the River Nile in Egypt, Winston Churchill is reported to have said: 'One day, every last drop of water which drains into the whole valley of the Nile ... shall be equally and amicably divided among the river people, and the Nile itself ... shall perish gloriously and never reach the sea' (quoted in McCully, 1996, p.18).

A similar attitude is revealed in these words from Joseph Stalin in 1929: 'Water which is allowed to enter the sea is wasted' (quoted in McCully, 1996, p.237).

Activity 4.2

What do these last two remarks reveal about the speakers' attitudes to the environment and their interpretation of the value of water as a resource?

Comment

Both statements suggest that the only consideration is the instrumental value of water as a resource for human consumption. Churchill did acknowledge the need to share the resource fairly among the people of the Nile valley but his prediction that the Nile should 'perish gloriously' reveals an attitude to the environment that we may now find hard to credit. Any aesthetic, ecological or non-instrumental value was simply not considered. Stalin's comment also reflects attitudes prevalent at the time: nature was to be tamed, resources to be exploited and rivers to be used with little or no thought for, or understanding of, the environmental consequences. This is not necessarily to criticize – environment was a word hardly used at that time and environmental impacts were barely on the agenda. Since then there has been a significant shift in attitudes to the Earth and its natural resources. Today awareness of environmental impacts is much greater, although increased awareness does not necessarily mean that the potential impacts are adequately accounted for in decision making, as the rest of this chapter will show.

2.2 Increasing demand

Differing perspectives on the value of water may give rise to contests, but a question of great importance when looking at causes of conflict is simply: is there enough water to go round? As with any other resource, if supply is plentiful and exceeds demand then conflict over access and availability is unlikely. But if the size and number of demands increase and they cannot be fully supplied then the likelihood of conflict will increase. For freshwater, the statistics show that global consumption increased steadily throughout the twentieth century and is expected to continue to rise (see Box 4.1 and Figure 4.2).

Box 4.1 Water withdrawal, water consumption and water use

Water withdrawal means water removed from a source and used for human needs. Not all water withdrawn is necessarily consumed. It may be returned to the original source, possibly with changes in the quantity or quality (e.g. water used for cooling in power stations).

water withdrawal

Withdrawal is not the same as **water consumption**, which means water withdrawn and used in a way that prevents its immediate reuse (e.g. immediate reuse is not possible because water has evaporated, been contaminated or incorporated into a product, including crops).

water consumption

The term **water use** may be applied to both withdrawal and consumption.

water use

Figure 4.2 shows several different projected estimates of withdrawal after 1995. Note that there is considerable variation but that the consensus is an increase in

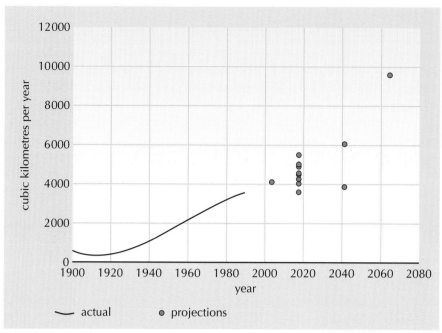

Figure 4.2 Actual and projected global fresh water withdrawals for the years 1900–2070.
Source: adapted from Gleick, 2000, p.43.

global water withdrawal through to the middle of this century. It is no surprise that this upward trend mirrors the pattern in world population over the same time period (Figure 4.3).

Activity 4.3

From Figures 4.2 and 4.3, compare the *rates* of increase of water withdrawal and population over the course of the twentieth century. How and why do they differ?

Comment

Reading from the graph in Figure 4.3, the world population of nearly 6,000 million in 1998 was approximately 3.5 times the 1900 population of 1,700 million. From Figure 4.2, water withdrawal increased from approximately 500 to 3,600 km^3/year over roughly the same period – an increase of more than seven times. In other words, water withdrawal has increased at more than twice the rate of population increase. This can only be explained by an overall increase in water use per head of population.

A combination of factors contributes to this trend. Some parts of the world have seen a general increase in living standards, which has led to a rise in the number of water-using domestic appliances such as washing machines, dishwashers and power showers. Over the same period, irrigation and industrial water use has expanded considerably all over the world. However, there are regional variations in these trends. Figure 4.3 shows that the predicted proportional increase in

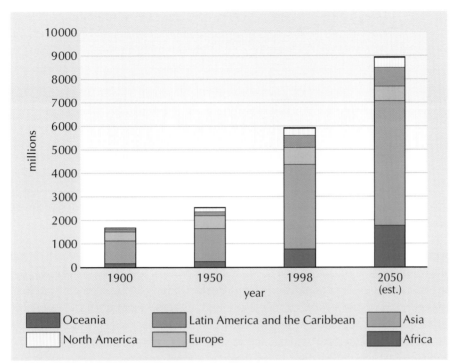

Figure 4.3 World population: actual and estimated in millions, by region, for the years 1900–2050.
Source: adapted from Gleick, 2000, p.67.

population of Asia and Africa by 2050 is much greater than that of other regions. Asia and Africa are also the regions that suffer from low and unreliable rainfall and where there are high levels of utilization of the total water resource. These differences contribute to significant inequalities in the provision of water supply and sanitation (Table 4.1).

Table 4.1   Access to water supply and sanitation, by region

Region	Population (millions)	Percentage of population with access		Number without access (millions)	
		water supply	*sanitation	water supply	*sanitation
Africa	784	62	60	302	318
Asia	3682	81	48	692	1916
Europe	728	96	92	25	55
Latin America and the Caribbean	519	83	76	89	127
North America	309	100	100	0	0
Oceania	30	87	93	4	2
Total	6052	82	60	1112	2418

* Sanitation encompasses all aspects of personal, household and public excreta and waste disposal. Data are for 1999.

Source: DFID, 2001, p.13.

In Table 4.1 the figures for regional differences mask the further inequalities in access to water supply *within* regions, where the distribution of both water and people is highly uneven. Exacerbated by the trend for rural people to migrate to urban centres, the greatest concentration of people occurs in large towns and cities where access to clean water can be the most problematic. Not only is there a high density of people needing water but also more waste has to be disposed of – usually into the nearest river. Living near a watercourse is no help if the water is so polluted it cannot be safely used. As Table 4.1 shows, over one billion people around the world lack access to a safe water supply and 2.4 billion lack adequate sanitation. As a consequence, millions of people in developing countries suffer from water-related diseases such as diarrhoea, malaria and schistosomiasis. Every day 6,000 people, mostly children under the age of five, die from diarrhoeal diseases (UNESCO, 2003).

This issue of inequality relates to the third key question for this book which is: can environmental inequalities be reduced? In terms of access to water supply, the statistics suggest that population levels are the underlying problem although, of course, there are economic, political and cultural factors that profoundly influence the provision of water and sanitation services as well. The uncertainties of climate change add another dimension to the problem. The scale of the fundamental, worldwide changes required to address inequalities such as these is vast indeed.

2.3 Not just for drinking

○   Apart from drinking it, what direct uses do human beings have for water?

●   Uses within the home include water for toilet flushing, washing (ourselves, clothes and dishes), cooking and possibly watering gardens. Non-domestic uses include commerce and industry, irrigation and other agricultural uses, power generation – both for cooling water and hydroelectric power. All these are direct uses of water that can be measured in terms of volume of water used per month or year.

In addition to direct uses, there are other, less easily measurable, water uses. For example, as mentioned in the previous section, we use rivers, lakes and oceans for diluting and dispersing large quantities of our waterborne wastes. This essential service carries with it the potential for considerable environmental damage if the water becomes polluted. Increasingly, we also make use of water for recreational and amenity purposes. Rivers, canals and lakes are popular venues for a wide range of leisure activities including fishing, sailing and other water sports. Many people enjoy just being near water for the aesthetic pleasure it gives, as mentioned in the quotations in Figure 4.1. And it is important to remember it is not only use by humans that should be considered. Water and wetlands provide many of the most important habitats on Earth for both plants and animals. The diversity of animals that live in, or depend on, aquatic environments is enormous, including not only all fish but also mammals, birds and amphibians as well as the less noticeable but innumerable invertebrates (such as molluscs, worms and insect larvae).

This wide range of water uses is made possible partly by the dynamic processes of the **hydrological cycle** . The downside of the interconnections of the cycle is that over-use for any one purpose, or any disruption of the cycle, can have serious knock-on effects elsewhere in the system. In the Thames valley in southern England, much of the water supply is obtained from groundwater reserves. During the 1970s and 1980s, demand for water was increasing and there were several years of unusually low rainfall. Pumping from the aquifers to meet the demand, in combination with reduced replenishment from rain, resulted in a widespread lowering of the water table across the Thames basin. The volume of water flowing in the rivers decreased and several tributaries dried up completely in some places during the summer months, causing acute damage to the river habitat on which so many plants and animals depend. Similar problems have affected several other rivers in southern England (Figure 4.4).

The **hydrological cycle** is the global movement of water, powered by the sun's energy, by which water evaporates from the oceans and land, condenses into clouds and is transported by winds, falls as precipitation and eventually returns to the oceans via the rivers. See **Blowers and Smith, (2003)** and **Brandon and Smith (2003)** for detailed discussion of the cycle.

Environmental damage of this sort can be avoided by taking an integrated and balanced approach to the management of the varying demands. Different uses can be accommodated as long as the needs of all are incorporated, and there is a will and capacity to do so. Grafham Water in Huntingdonshire, England, provides an example of where different uses are successfully combined. Grafham Water was built as a water supply reservoir, but it also provides a valuable habitat for wildlife and is used for many recreational purposes (sailing, windsurfing, fishing and bird-watching). Zoning of the reservoir for these different activities allows them to be integrated with the primary function of water storage (Figure 4.5), thereby reducing the potential for conflict among the users.

Figure 4.4 River Wylye in Wiltshire, England: (a) in summer and (b) in winter. River flow naturally varies with the seasons but water abstraction was reducing summer flow to a trickle rather than the full flow of winter. In an ongoing programme of improvement, abstraction management in the upper reaches of the river has been altered so that higher flows are maintained throughout the year.

In the context of water management, taking an integrated approach can provide one answer to the second chapter question: how can potential conflict over water be avoided? Recognizing and incorporating the needs of different interest groups can improve their mutual understanding and thereby reduce the likelihood of conflict among users. Unfortunately, there are many examples where the multiple uses of a river system, especially the wider

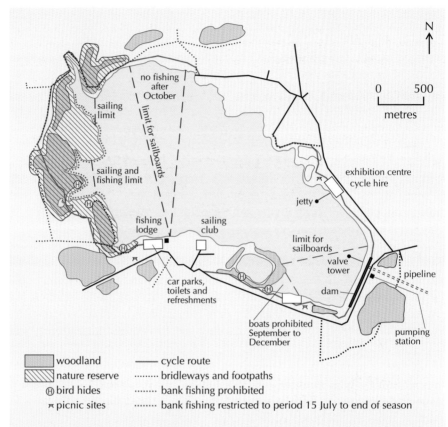

Grafham Water is a water supply reservoir covering 600 hectares near Huntingdon. It has been zoned to allow the integration of wildlife conservation and several different leisure activities with the primary function of the reservoir, which is to store water prior to treatment and distribution. Many people visit the reservoir for windsurfing, sailing, fly-fishing, cycling, nature trails, birdwatching or simply to enjoy a picnic by the water's edge. Three car parks, each with toilets, refreshments and a children's play area, provide the necessary facilities for the numerous visitors. The main part of the reservoir is available to all users. In two smaller zones, sailboards are prohibited but sailing and boat fishing are permitted. The sheltered bays at the western end are wildlife sanctuaries where no windsurfing or sailing is allowed. In the more southerly of the two, fishing boats and bank fishing are also prohibited to provide an area completely free of disturbance. Access to this area from land is carefully controlled by clearly way-marking footpaths, bridleways and cycle routes, in order to protect the wildlife further.
Source: adapted from Anglian Water plc publicity leaflets.

Figure 4.5 Reservoir zoning at Grafham Water.
Source: Furniss and Lane, 1990.

environmental use, are not all taken into account when decisions are made. In these circumstances, contest over environmental issues will arise, as can be seen in the story of the Alqueva Dam.

2.4 Case study: Alqueva Dam

On the Guadiana River at Alqueva, in southern Portugal, a controversial dam project is underway that will create one of the largest artificial lakes in Europe, with a planned area of 250 km^2 (Figures 4.6 and 4.9). In February 2002, a ceremony at the site celebrated the closing of the 96 metre high floodgates and the reservoir began to fill. The occasion was also marked by campaigners who attended the ceremony to protest against the dam.

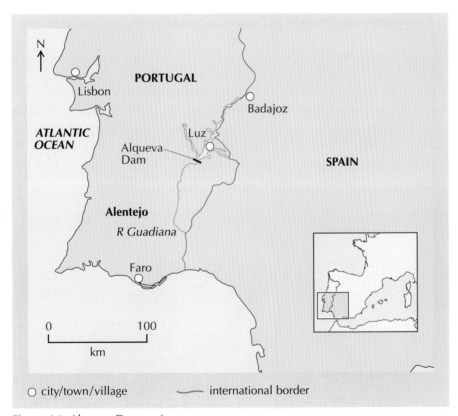

Figure 4.6 Alqueva Dam region.

Conservation campaigners complained that vegetation, including more than one million trees, had been cut down over 135 km^2 to make way for the reservoir. The area is an important habitat for wildlife and is inhabited by many rare and protected species, including the black stork (*Ciconia nigra*), the endangered Iberian lynx (*Lynx pardellus*), shown in Figure 4.7, and the Iberian imperial eagle (*Aquila adalberti*), shown in Figure 4.8. EU laws prohibit development in areas that can interfere with the habitat of the Iberian lynx, a priority protected species.

environmental impact
assessment (EIA)

Conservation groups argue that the project contravened the EU Habitats Directive and criticized the **environmental impact assessment (EIA)** of the project as inadequate (see Box 4.2). Among several failings, the EIA did not identify the lynx as under threat and did not consider the possible alternatives to the scheme.

The village of Luz (see Figure 4.6) and its 400 inhabitants have been moved to make way for the reservoir. In a bizarre scheme to accommodate the displaced villagers a replica of Luz has been built a few kilometres away at a cost of £20 million, with replacement homes next to the same neighbours. The cemetery has been moved with them. In the words of one villager (reported in Wilkinson, 2002): 'People have now begun to realize that the moment when we have to change has come and it is very, very sad. This grief cannot be paid with any money.'

The Portuguese government and other supporters of the dam argue that the reservoir will regulate flow in the lower reaches of the river and will allow Portugal to take advantage of the waters of a river that runs through its territory and which have not been fully exploited. The water in the reservoir is intended for irrigation of about 110,000 hectares in the Portuguese region of Alentejo, but campaigners against the dam point out that the area's existing irrigation network is used at less than 50 per cent capacity. A hydroelectric power station with capacity to supply 180,000 people will also be built.

Figure 4.7 The Iberian lynx (*Lynx pardinus*): less than 800 remain, in pockets in Spain and Portugal.

Figure 4.8 The Iberian imperial eagle (*Aquila heliaca*): the world population of only some 150 breeding pairs is confined to Iberia.

Box 4.2 Environmental impact assessment (EIA)

The purpose of an EIA is to 'predict, analyse and evaluate the impacts of a proposed action on the environment and ensure that information regarding these impacts is taken into account in decision making' (Roberts, 1995)

EIAs became part of European law in 1985 and were adopted as standard practice in most parts of the world by the end of the twentieth century. Ideally, they should involve four stages:

- *Starting up*
  This stage asks whether an EIA is needed for the project, what impacts and issues should be considered, and what alternatives exist.

- *Impact prediction, evaluation and mitigation*
  The likely environmental impacts are predicted, their potential seriousness evaluated, and the possible measures to reduce or remove environmental impacts are investigated.

- *Participation, presentation and review*
  This stage includes public consultation and review, which then feeds into the presentation of an environmental impact statement, which is then reviewed and considered as part of the decision-making process.

- *Monitoring and auditing*
  After the decision is taken, the consequences and effects should be tracked and assessed.

(adapted from Blackmore and Ison, 1998, p.90)

Even within a framework like this, there is room for interpretation and selective inclusion of the components. Public consultation may be minimal or non-existent. What constitutes the 'environment' may not be consistent. For example, impact on archaeological sites was not initially included in the assessment of environmental impact at Alqueva. The displacement of local people, who for some big dam projects number many thousands, has not always been included in an EIA and it has been argued that displacement is a social, rather than environmental, impact. The independence of the assessors may also be called into question, especially in instances where the EIA has been prepared by the company contracted to work on the project itself. The quality and robustness of EIAs depend ultimately on the assessing team, their independence and their personal assumptions and values.

The project has been plagued by problems including the discovery that the main dam wall has been built on a double seismic fault line making it potentially prone to earthquake damage. Another battle arose when irreplaceable prehistoric rock art engravings were discovered in the area to be submerged by the reservoir. The engravings have been hailed as 'one of the most important archaeological discoveries in recent times' (ENS, 2001b). The Portuguese Prime Minister said that the importance of Alqueva Dam (Figure 4.9) to the national economy was such that work would go ahead regardless of the engravings (ENS, 2001b).

Figure 4.9 Alqueva dam during construction, March 2001.

In addition, the development company have been promoting the new dam as a tourist facility. Dozens of applications have been submitted to local authorities by multinational corporations to build new golf courses, casinos, health spas, private resorts and water sports facilities in and around the new lake.

Environmentalists, meanwhile, warn that reduced flow below the dam will increase pollution and have a negative impact on the biological productivity of the river downstream, one of the last wild woodland valleys in Europe. A lynx conservation group commented:

> This area is one of the most ecologically rich in Europe. Official reports show that several species of animals, plants and fish found nowhere else in the world will be lost forever because of this dam. It is also one of the best lynx habitats still left. It is incredible that when such a unique species as the Iberian Lynx is teetering on the brink of extinction, that one of its prime habitats should be deliberately destroyed.
>
> (ENS, 2001b).

Activity 4.4

From what you have just read, identify the planned benefits and the damaging consequences of the Alqueva Dam.

Comment

Planned benefits are: river flow regulation, irrigation, electricity generation, tourism and general economic gain.

Damaging consequences are: loss of habitat for lynx and other wildlife, loss of prehistoric engravings, relocation of Luz, possible pollution and reduced downstream productivity.

The conflict of interest arising from the different uses of the water is clear. The government and supporters of the dam wanted to take advantage of the economic benefits identified above, whereas the opponents put greater value on the environmental, conservation and archaeological losses.

Summary

This section began with consideration of the key concept of value, as it applies to water. The value given to water by us as individuals will depend on the availability of water, our location and our economic status. As well as the importance to us as direct water users, water is also valued for its role in environmental systems throughout the world.

Two of the book's key questions have been considered in relation to water and conflict about water:

- *How and why do contests over water arise?*
  Two main reasons have been discussed. The first is that the global demand for water is going up as a consequence of rising population levels as well as increased use per person. Increased demand for any finite resource is going to be a potential cause of conflict. The second reason is that we use water for many different purposes, both directly and indirectly, but these uses may not be mutually compatible and can lead to conflict between the different interest groups, as illustrated by the Alqueva case study.
- *How can potential water conflict be avoided?*
  This question relates to environmental understandings, and in this section we have looked at how it can be possible to accommodate opposing demands if an integrated management approach is adopted that takes all uses into account.

As to the third question, while it is apparent that there are great inequalities in access to water and sanitation throughout the world, the question of how they might be reduced has no easy answers.

3   Upstream to downstream: actions and impacts

In this section we will look further at the first chapter question and reasons for disputes between water users and, in order to answer the second chapter question, at how an appreciation of the complexity of a system can aid understanding.

3.1 Whose water is it anyway?

The idea that someone owns the water flowing in a river is somehow at odds with its position in the hydrological cycle. (Who owns the clouds?) There is a widely held view that water is a common good and should not be privately owned. Yet the ownership of rivers and rights to water resources is at the root of many water conflicts. In many parts of the world, rivers and the water flowing in them are claimed by the owners of the river banks (see Box 4.3). These *riparian* owners (Latin: *riparius* = bank) own the stretch of river on their land although their use of the water may be limited by regulation or licence. Whichever system of rights applies, the notion that someone owns the water does not take account of the fact that, as water users, we are all **stakeholders** in the water supply system.

A **stakeholder** is any individual or group of people who share an interest or stake in a particular issue or system (e.g. an organization, a campaign, a project).

Box 4.3 Water rights

The three major systems of water rights are:

- *Riparian rights*
  Ownership, or reasonable use, of water is linked to ownership of the adjacent or overlying (of aquifers) lands. Riparian rights are derived from Common Law as developed in England and, as a consequence, this principle is mainly found in the UK and in former British colonial countries.

- *Public allocation*
  This involves administered allocation of water rights and seems to occur mainly in so-called 'civil law' countries i.e. that derive their legal system from the Napoleonic Code such as France, Italy, Spain, Portugal, the Netherlands and their former colonies.

- *Prior or historical rights*
  These rights are based on the appropriation doctrine under which the water right is acquired by actual use over time. In other words, if you have made use of the water supply in the past, then you have acquired the right to continue to use it. This system developed in the western part of the USA.

(adapted from Savenije and van der Zaag, 2000)

Another difficulty with the concept of owning river water is that it does not stay in one place but flows from source to mouth. The upstream riparian owner may wish to exert their right to abstract water from 'their' river, either to use immediately or to store for future use or they may wish to discharge wastewater into it. If, in so doing, they significantly reduce the river flow or pollute it, then the downstream owner, who also has a claim on the water, will suffer.

A **catchment area** or basin is the total land area drained by a river system i.e. a main river and all its tributaries.

Rivers do not neatly follow the same boundaries that humans have established for their territorial claims so there is unlikely to be any correlation between the pattern of land ownership and the **catchment area** of a river. Typically, any river will have several riparian owners over its course and so there is clearly considerable potential for argument over opposing claims. The UK has an advantage as an island in that international disputes over ownership, discussed later in this chapter, are not an issue but on all scales (local, regional or international) the connection between upstream activity and downstream impact is inescapable.

In England and Wales, riparian rights are overlain by a water management structure consisting of many different organizations and agencies, each responsible for different aspects of water services. For example, the water companies provide water supplies and sewage treatment, the Drinking Water Inspectorate monitors the quality of water supplies, the Office of Water Services (OFWAT) regulates the price of water, and the Environment Agency controls abstraction from rivers and monitors river water quality. Each organization has its own area of responsibility but inevitably they are closely interrelated. Complex, interconnected situations like this can sometimes be better understood by thinking of them as a **system** (see Box 4.4).   system

Within the system of water services in the UK, other stakeholders are the various categories of water users as described in the previous section. One useful way of representing a complex system like this and getting to grips with the various components and their relationships is by using a systems map (see Box 4.4 and Figure 4.10).

Box 4.4 Systems and systems maps

A *system* can be defined as a selected set of related components together with the connections between them, or more eloquently, as an integrated whole whose essential properties arise from the relationships between its parts. The selection of the components is one that someone chooses for a defined purpose and, in that sense, system should be regarded as a concept, not necessarily something that has an independent existence. The person defining the system would also choose what was part of the system and what was not i.e. they would choose a notional boundary around it. [See also **Morris, 2003**.]

Systems maps are a type of diagram. Drawing a systems map can often help to develop your understanding of a situation, i.e. your *system of interest*. They can be used either to group or to link components of a system together to show either structure or process or both. Each element or subsystem is contained in a circle or oval. A line, the *system boundary*, is drawn round a group of elements or sub-systems to show that the things outside the line are part of the environment, while those inside the line are part of the system.

[You can see an example of this format in Figure 4.10.]

(adapted from Lane et al, 1999, p.65)

For the water industry, the main components of the system of interest are the organizations, agencies and other stakeholders. Figure 4.11 is a map of the system. In it, I have chosen to group the elements in three subsystems: providers, regulators and users.

The title of a systems map is very important – it should be accurate and acknowledge any known limitations. For example, my map applies only to England and Wales because the management structure is different elsewhere in

Systems maps consist of blobs of varying sizes, words and a title. The conventions for drawing a systems map are:

- The lines around the blobs (1–6) represent boundaries of system components.

- Words (e.g. aaa, bbb, ccc) are used to name each system or component.

- Blobs (5 and 6) outside the main system boundary (1) represent components of the environment.

- Blobs (2, 3 and 4) inside the system boundary represent components of the system. Components (e.g. 3) can be shown as grouped into sub-systems (2).

- Blobs may overlap but only if some components (which need not be depicted) are seen as common to both in the early stages of identifying a system of interest.

- A title defining the system of interest is essential.

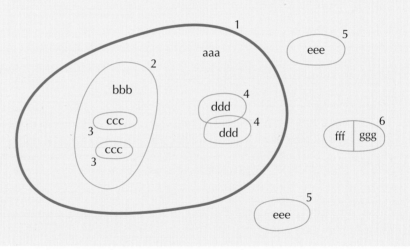

Figure 4.10 Format for a systems map.
Source: adapted from Lane et al, 1999, p.65.

the UK, and indeed in the rest of the world. My first attempt at writing the title for the map was to label it 'The water management system of England and Wales'. Then I realized that it extended beyond management because it included the users, so I amended it. Also it only refers to stakeholder groups and not any wider systems such as infrastructure (reservoirs, treatment works), the regulatory environment (Acts of Parliament, EC Directives) or hydrological environment (river abstraction, aquifers). A different systems map could be drawn that included these elements of the system environment as well as the stakeholder system. My final title therefore reflects these considerations. It is always important to remember that any systems map is only a representation of the view of the person(s) who drew it and it will be influenced by their values and priorities – other perspectives may be equally valid.

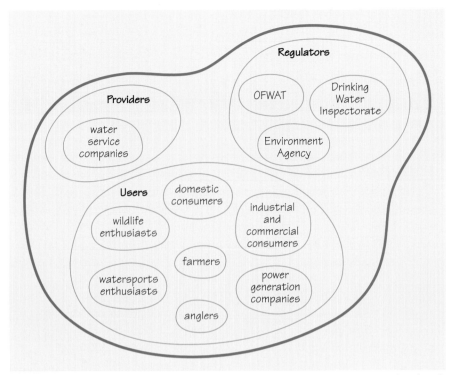

Figure 4.11 Systems map of the stakeholder groups of water services in England and Wales.

The advantage of a long established and thoroughly regulated system like that in the UK is that conflict between owners and other stakeholders can (usually) be avoided. This can be achieved by recognizing the complexities of a system and the interrelationships between the component groups. The following case study illustrates a complex stakeholder system, the components of which are linked to the multiple demands placed on a river in a situation where water resources are limited and demand is high. It provides a further example to demonstrate how disputes over competing water uses can arise. The case study also relates to two key concepts of this book, power and action, by showing how different resources of power can be deployed by the actions of interest groups.

3.2 Case study: Okavango River

The Okavango River in southern Africa rises in the highlands of Angola, flows southwards forming part of the border between Angola and Namibia and continues into Botswana where the river disperses in a low-lying area to the north of the Kalahari Desert. Unusually, the river does not flow to the sea but is landlocked and the waters fan out into numerous channels that together form the Okavango Delta (Figures 4.12 and 4.13).

This is the largest freshwater delta in the world, covering an area of approximately 15,000 km^2. Its ever-changing pattern of grassland, intermittent floods and swamps creates a rare combination of habitats that makes it a haven for wildlife (Figure 4.13). The rich and varied fauna includes several endangered species. Tens of thousands of people also live in the region and depend on the waters of the Okavango for their livelihood, either directly or through the growing ecotourist industry. Seasonal flooding is an essential part of the ecosystem. Following the rainy season in the highlands, each year the Okavango floods and stretches across thousands of square kilometres transforming the parched landscape into a verdant oasis.

In the 1980s a major development scheme was proposed by the Department of Water Affairs (DWA) of the Botswana government, called the Southern Okavango Integrated Water Development Project (SOIWDP). The purpose of the project was for 'optimum use of land and water resources culminating in increased agricultural production benefiting the region and the nation' (quoted in Thomas et al., 2001, p.109). In the view of some politicians and engineers, the dispersal through the delta and eventual disappearance of almost all of the water was a waste. The plan for the SOIWDP was to dredge one of the main river channels through the delta and construct three dams to store the increased flow. Each of the resulting reservoirs had a different primary purpose. The first was to provide water for the rapidly growing village of Maun. The second reservoir was intended for agricultural use and the third was to benefit the diamond mine at Orapa, the diamond industry being a vital part of Botswana's economy.

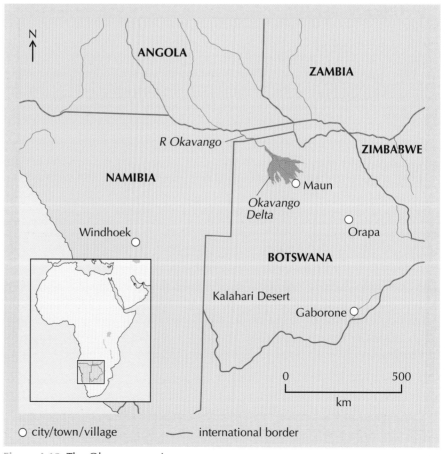

Figure 4.12 The Okavango region.

Figure 4.13 Red lechwe (*Kobus leche*), a rare antelope, with other wildlife in the Okavango delta.

Initially, local people and district officials accepted the scheme – the predicted improvement in water supply was much needed. The Kalahari Conservation Society (KCS), a Botswana environmental group, expressed some concern over the proposed dredging, but the details of the scheme were not known in the early stages so protest was limited. The DWA consulted with other interested parties about the project. The Okavango Water Development Committee (OWDC) had been established as a forum for discussion about water issues in the region. By the late 1980s the Committee had representatives from other government departments, the engineering firm that designed the project and the KCS. Botswana has a strong democratic tradition and the public was also consulted through traditional village meetings called *kgotlas*, organized to discuss the project. The term 'river improvement' was used in these consultations, giving the impression of limited work. Dredging was not mentioned, so the local people did not envisage a large-scale project. These local meetings took place at the same time as the design and details of the scheme were being prepared. In 1988 the cabinet approved implementation of the first stage and in late 1990 construction started.

Activity 4.5

What were the existing uses of the river and the intended uses for the river under the SOIWDP scheme?

Comment

The planned purpose of the project was for water supply for the inhabitants of Maun, for agricultural use and for use by the diamond mine. A continuing use of the water was to sustain the delta ecosystem downstream. There were two aspects to this – both use of water to maintain the intrinsic ecological value of the delta and its wildlife and the economic use of the delta as the main attraction for the tourist industry.

The unexpected arrival of heavy dredging machinery on site prompted action. The local residents called meetings and organized themselves into an action group that later became the Tshomerolo Okavango Conservation Trust (TOCT). The KCS did not take an active role in the opposition campaign; they believed it might damage their credibility with government if they publicly opposed official policy. In January 1991 the local community made their feelings known at a *kgotla* lasting many hours and attended by several hundred people. The government minister present was questioned by at least twenty speakers, none of whom supported the project.

In addition to the local opposition, there were reports in the press that, if the project went ahead, Greenpeace International had threatened to call for a worldwide boycott of diamonds under the emotive slogan 'Diamonds are for Death'. Greenpeace sent a visiting team who appealed to the government of Botswana to establish the Okavango delta as a World Heritage Site. The team's report was widely publicized in the international press. In fact the owners of the diamond mine were not strong supporters of the scheme – they had alternative water sources available to them and dropped out of the picture as the opposition grew.

In response to the mounting objections, the Botswana government called a temporary halt and commissioned an independent review of the scheme by the International Union for the Conservation of Nature (IUCN). The recommendation of the IUCN team was that the project should be terminated. They argued that insufficient attention had been paid to alternatives and that there had been inadequate analysis of the impact on the environment. They concluded that the project would have an adverse impact on the environment, economy and tourism potential of the region as a whole. In May 1992, as a result of the local opposition, the IUCN study and the potentially high economic and environmental costs, the government announced that they would not be proceeding with the project.

Activity 4.6

According to the account you have just read of the Okavango river project, who were the stakeholders in the SOIWDP? Try to group your list into appropriate sub-systems. Draw a systems map of the stakeholders.

Comment

The stakeholders mentioned were: the local inhabitants, including Maun residents; local farmers and other local people whose livelihood depended on the river and/or its associated tourist industry; the Botswana government and its Department of Water Affairs; the Kalahari Conservation Society; TOCT; OWDC; the diamond mine company; and later on, Greenpeace International and IUCN.

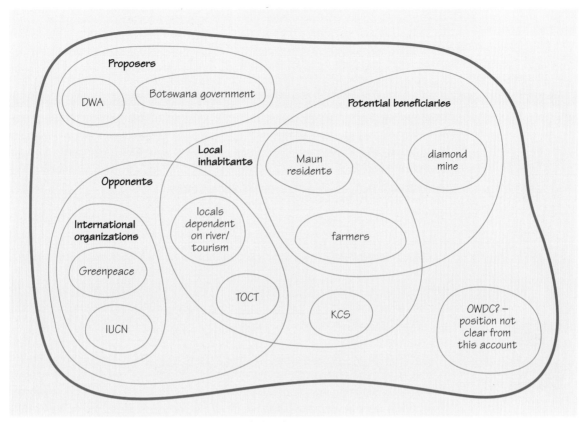

Figure 4.14 A systems map of the SOIWDP stakeholder system.

From this list, the local inhabitants could be grouped as one subsystem. Other subsystems emerge from reading the account – some of the stakeholders were opponents of the scheme and another set were potential beneficiaries. The opponents included the two international organizations, IUCN and Greenpeace, as well as some of the local groups. The potential beneficiaries were the diamond mine and also some of the locals so there are overlaps between subsystems here. The government and DWA appear to be a subsystem on their own. My representation of the system is shown in Figure 4.14. Do not worry if your map doesn't look exactly like this – yours is based on your own interpretation of the case study story as it is told here, which may differ from mine.

The Okavango story illustrates how activities in one part of a river system have consequences elsewhere and it also demonstrates how the multiple uses of water lead to a complex system of stakeholders associated with those different uses. Each of those stakeholders had different resources of power and deployed them in different ways. The Botswana government had power over the whole project because they had decision-making authority on whether to proceed with the scheme or not. The DWA had power to influence that decision in their role as advisors to the government. It has been argued that the project served an additional purpose for the DWA, which was to increase its prestige and demonstrate its own power within the state's bureaucracy. Abrams Neme (1997, p.42) describes the DWA position: 'Its responsibility for the nation's water gave it a measure of control over all sectors of the economy. By working to ensure universal availability, it justified its own *raison d'etre* and improved its status within the public service.'

The external agencies also had power to affect the outcome. IUCN had power through their expertise and prestige and they were able to exert influence on the decision in their role as experts. Greenpeace had power derived from the publicity they could get to bring criticism on the Botswana government if the government had decided in favour of the scheme.

When the scheme was first proposed, the local villagers felt powerless. They were not included as stakeholders in the process. It was said that 'there was no need to discuss projects like this because everything was already discussed, finalized and decided upon in [the capital] Gaborone' (Abrams Neme, 1997, p.47). But they were not as powerless as it seemed at first and, later on, when the realities of the project were recognized the local groups demonstrated their power by taking action and protesting against the project.

One advantage of drawing a systems map, like the one in Figure 4.14, is that it can reveal aspects of the situation that were not previously apparent. The idea of power deriving from the combination of resources (discussed in Chapter One and **Hinchliffe and Belshaw, 2003**) is clearly demonstrated in Figure 4.14 in the subsystem of 'opponents'. The combined resources of organizations and local people contributed to the defeat of the scheme although quite how these different interests affected the decision making is not fully known. It has been argued that the government's obligation to its democracy, along with the threat to its diamond industry, forced it to shelve the scheme. More directly, as argued by the opponents, the possible impact on the delta downstream was such that the disputed benefits from the scheme could not be justified. An added factor may have been that there was not only an environmental cost if the delta was damaged but also possible actual economic cost to the tourist industry. In this instance it appears that a fortunate coincidence of economic and environmental interests conspired to defeat the scheme.

3.3 Big dams

At the centre of the case studies described so far in this chapter have been plans to build a dam. In both stories the focus was on the consequences of damming a river, with profound differences of opinion between the conflicting interest groups as to the pros and cons. These two stories are not unusual.

No examination of contemporary water conflicts would be complete without giving attention to the disputes associated with large dam projects around the world. Big dams are one of the most controversial large-scale development projects of the past 100 years. In 1900 there were no dams in the world higher

than 15 metres (the conventional definition of a large dam). By 1950 there were
5,270. By the end of the twentieth century there were more than 40,000 – about
half of which are in China. 'Large' dams are dwarfed by the 'major' dams, which
are defined as those higher than 150 metres. In 1950 there were 10 major dams in
the world, including the first, Hoover Dam, which was the world's highest dam for
more than two decades until 1957. By 1995 there were more than 300 dams of this
size (McCully, 1996). Since 1980 the development of new dams has slowed in
some parts of the world as awareness of their environmental and social impact
began to redress the balance against their advantages. In the USA, several dams
have been removed or are scheduled for removal, but in most parts of the world
the impetus is still firmly in their favour, the Alqueva being just one of the more
recent examples.

It is undeniable that building a dam offers the opportunity to affect profoundly the
economic prosperity of vast areas of land. Improved power, agriculture, water
supply and recreation opportunities are all enticing prospects that could
contribute to the productivity of a region. Flood control is a major issue on
many rivers where annual flooding can cause loss of life as well as damage to
property, infrastructure and productivity. The incentives for dam building are not
hard to see but there are costs – environmental, social and financial.

The potential for conflict arises from the contrast between the advantages and
disadvantages of dams. Those in favour of dams would argue that the benefits
outweigh the costs but the anti-dam campaigners say that the impacts are
unacceptably great and that any benefits are short term and usually overestimated.
Frequently the environmental impact is underestimated or ignored. Even with the
general acceptance of EIAs as standard practice for large projects, the
assessments on dam schemes have often been criticized as biased or inadequate,
for example by omitting to include impact on the local people. As well as the loss of
land and livelihood, the human impact may include increased health risk from the
presence of permanent water in the reservoir and in irrigation channels. This
provides more opportunities for waterborne diseases such as schistosomiasis,
diarrhoeal diseases, malaria and other mosquito-carried diseases.

In these situations, great power is held by the decision makers who approve the
schemes. Part of the appeal of building a big dam probably lies in aggrandize-
ment, attainable – both collectively and personally – by the organization or
government or associated individuals responsible for it. Theodore Steinberg, a
US historian, said of the Hoover Dam: '[It] was supposed to signify greatness,
power and domination. It was planned that way' (quoted in McCully, 1996, p.3).
But it is the environmental interests and the local inhabitants – who are relatively
powerless – on whom the impacts of dam building are likely to fall.

Clearly, larger schemes are going to affect more people and it is the enormous
scale of modern dams that has led to the numerous protests from environmental
groups against dam projects all over the world. The Three Gorges Dam on the
River Yangtze in China, when completed, will be the largest dam in the world

Figure 4.15 Three Gorges Dam under construction, Yichang, July 2001; when complete it will be the largest dam in the world.

(Figure 4.15). It will be as high as a 40-storey building and the reservoir will extend 660 km up the valley behind it. The most significant benefit expected from the scheme is the control of flooding of the Yangtze, which historically has claimed millions of lives, although some have questioned whether the dam will be effective. On the other hand, the forecast statistics on the impacts of the project are staggering. The scheme will displace an estimated 1.3 million people; some estimates go up to 1.9 million. When full, the reservoir will submerge 137 cities and towns, 1,300 factories, 1,100 villages and 178 rubbish dumps (Rennie, 2002).

Protests against the Three Gorges Dam project have been ultimately ineffective but active campaigns against many other dam projects continue around the world, as is evident from the many reports in the international press.

Summary

In this section, some further answers to how and why contests over environmental issues arise have been explored. The constant movement of water through the hydrological cycle means that activities affecting one part of the cycle tend to have knock-on effects elsewhere. This is especially relevant in rivers where actions by upstream users will have an impact on the quality and quantity of water downstream. The borders of river systems do not correspond to human-devised boundaries and this can lead to conflicting claims of ownership. The notion of ownership is not always helpful when a more realistic picture would be a complex system of stakeholders, reflecting the multiple uses of water, who all share an interest in a water resource. Systems maps have been introduced as a technique for describing and clarifying complex situations.

The Okavango case study showed how actions in one part of a river could have had impacts elsewhere in the system. Aspects of the story relate to all three key concepts used in this book: stakeholders *valued* the river and its water for different reasons, held a range of *power* positions and took different *actions* that ultimately led to the scheme being shelved.

Most, if not all, of the causes of water conflicts are demonstrated by big dam projects. Increased demand for water resources, multiple uses and stakeholders, downstream impacts arising from upstream action, river systems crossing boundaries and ownership rights are all to be found in such projects.

4   Water wars – myth or reality?

So far, this chapter has concentrated mostly on water issues at the local and regional scales. International rivers provide a context for further exploring the key question: how and why do contests arise? When rivers cross borders between neighbouring nation states, the potential for conflict arising from disputed rights and competing demands for water can only increase.

4.1 Sharing international rivers

Many rivers, lakes and groundwater aquifers are shared by more than one country and there are many instances of rivers that flow through several different countries (Table 4.2 and Figure 4.16). Aquifers present an added difficulty in that the geographical extent of the underground reservoir may be uncertain.

Table 4.2   International river basins shared by eight or more states

River basin	No. of states	States sharing the basin
Danube	17	Romania, Hungary, Yugoslavia (Serbia and Montenegro), Austria, Germany, Bulgaria, Slovakia, Bosnia-Herzegovina, Croatia, Ukraine, Czech Republic, Slovenia, Moldova, Switzerland, Italy, Poland, Albania
Congo	11	Democratic Republic of Congo, Central African Republic, Angola, Republic of the Congo, Zambia, United Republic of Tanzania, Cameroon, Burundi, Rwanda, Gabon, Malawi
Niger	11	Nigeria, Mali, Niger, Algeria, Guinea, Cameroon, Burkina Faso, Benin, Ivory Coast, Chad, Sierra Leone
Nile	10	Sudan, Ethiopia, Egypt, Uganda, United Republic of Tanzania, Kenya, Democratic Republic of Congo, Rwanda, Burundi, Eritrea
Rhine	9	Germany, Switzerland, France, Netherlands, Belgium, Luxembourg, Austria, Liechtenstein, Italy.
Zambezi	9	Zambia, Angola, Zimbabwe, Mozambique, Malawi, United Republic of Tanzania, Botswana, Namibia, Democratic Republic of Congo
Amazon	8	Brazil, Peru, Bolivia, Colombia, Ecuador, Venezuela, Guyana, Suriname
Lake Chad (internal drainage)	8	Chad, Niger, Central African Republic, Nigeria, Algeria, Sudan, Cameroon, Libya

Source: Gleick, 2000, p.34.

Generally speaking, the upstream riparians have the advantage over the downstream countries in that they get the water first. However, the existence of international shared water basins does not inevitably lead to conflict between the neighbouring countries. Certain geographical and geopolitical circumstances influence the likelihood of water becoming a source of confrontation. The degree of water scarcity that already exists in the region, and the availability of alternative sources, are bound to be central to the issue – as noted earlier, disputes are unlikely where water is plentiful. Water quality as well as quantity may be the source of disagreement so the degree of water pollution in a shared river is also relevant. The power relationships and the cultural, political and economic links between water-sharing states are critical.

To reduce the possibility of conflict arising over water, or any other shared resource, the countries concerned must reach agreement on the principles by which they will share. In the historical context of water use, two extremes have emerged. The first is the principle of *total sovereignty*. This assumes that each

Figure 4.16 International river basins in Europe; note how the borders of the river basins do not coincide with the state boundaries.

state has complete sovereignty over the drainage basin, or part of the basin, located within its territory and does not allow any other claims on the resource. This is also known as the Harmon Doctrine, after Judson Harmon who was US Attorney General in 1895. In a dispute with Mexico over water rights on the Rio Grande, he stated that there was no reason why the USA should take Mexico's needs into consideration. This argument has tended to be favoured by upstream countries as they can then make whatever use of the water they choose without consideration of downstream users. Downstream states may take the position that they have acquired rights to the quantity and quality of the water they have used in the past and so claim historical precedence. At the opposite end of the spectrum to the Harmon Doctrine is the concept of *community of interest*. This principle holds that within a shared river basin, no single riparian can block the action of any other; in other words, any use of the water resource must be agreed by all other riparian states. The middle ground between these two extremes is the principle of *limited sovereignty* which, in this context, can be summarized as 'use what is yours so as not to cause harm to another' (Waterbury, 1993, p.47).

Several attempts have been made to codify the principles for international agreements. The Helsinki Rules on the Uses of the Waters of International Rivers, drawn up in 1966, specified eleven principles to guide water use. Principles based on the Helsinki Rules have since been developed further and were adopted by the United Nations in 1997 in the Convention on the Law of Non-Navigational Uses of International Watercourses. The central principles of the UN Convention are:

- Watercourse states shall utilize an international watercourse in an equitable and reasonable manner. Watercourse states have both the *right* to utilize the watercourse and the *duty* to cooperate in the protection and development thereof.
- Watercourse states shall, in utilizing an international watercourse, take all appropriate measures to prevent the causing of significant harm to other watercourse states.

In practice the difficulties arise with judging and reaching agreement on what constitutes 'equitable and reasonable' use and 'significant harm'. These are open to interpretation by the states involved.

Cooperation and mutual consideration of the needs of others are essential in reaching international agreements. Such an agreement has been made concerning the River Danube. The Danube flows through 17 states in eastern Europe (see Table 4.2 and Figure 4.16) and has suffered pollution from urban and industrial waste and agricultural run-off along its course for decades. Over the last century, 80 per cent of the Danube's wetlands and floodplains have been destroyed by the combined effects of pollution and dam building and other development along the river. In May 2001, an agreement on international cooperation for environmental protection was signed by representatives of 15 countries. The Romanian president said at the time,

> Some of the daunting problems concerning the rational use, ecological rehabilitation and adequate protection of the Danube go way beyond the possibilities of individual countries ... Water pollution, flood control and other issues with a significant environmental impact are regional in nature and they have to be addressed at a regional level with adequate international guidance and support.
>
> (ENS, 2001a)

4.2 History of water conflict

Cooperation of the sort demonstrated over the Danube is not always achieved. In many situations there are conflicts of interest between neighbouring countries, unrelated to water resources, but concerning other political, ideological or territorial differences. In such circumstances disputes over water can be a symptom, rather than the cause, of conflict. Gleick has prepared a history of water conflict that dates back at least 5,000 years (Table 4.3 shows some recent entries).

Of all the incidents that could be described as water conflicts, many are within, rather than between, nations and have more to do with internal political struggles than trans-border disputes. Very few are solely about water resources where water supplies or access to water are at the root of the tensions. Some are development disputes where water resources are a major source of contention in the context of economic and social development. In many cases, water or water systems are used as military or political tools by both state and non-state actors.

What then is the answer to the question: are water wars a myth or reality? Despite the lengthy chronicle of water conflicts, none of these have, as yet, developed into a water war, where this is defined as a violent conflict that is directly caused by the incompatible sharing of water resources. There are opposing views on the likelihood of a water war. Some argue that water scarcity, increasing demand and global climate change will inexorably lead to growing tension and ultimately to armed conflict between states. Others argue that there is likely to be more than one motive for starting a war, and that water is only one among several issues, including ethnic antagonism, ideology, border disputes and religion.

Whether a dispute leads to violent conflict or not, the situation can be ameliorated if nations reach an understanding and can agree to cooperate over equitable sharing of water resources. However, some complex situations, as exemplified by the Tigris–Euphrates story that follows, appear to put very considerable demands on the principles of cooperation and agreement. This final case study again addresses the key question: how and why do contests over water arise? In this case the context is international dispute over rights to water resources in a situation where there is unequal distribution of those resources between neighbouring countries.

Table 4.3   Water conflict chronology – examples from mid 1970s onwards

Date	Parties involved	Description
1974	Iraq, Syria	Iraq threatens to bomb the dam at al-Thawra in Syria and masses troops along the border, alleging that the dam has reduced the flow of Euphrates River water to Iraq.
1975	Iraq, Syria	As upstream dams are filled during a low flow year on the Euphrates, Iraq claims that flow reaching its territory is 'intolerable' and asks the Arab League to intervene. Syrians claim they are receiving less than half the river's normal flow and pull out of an Arab League technical committee formed to mediate the conflict. In May, Syria and Iraq reportedly transfer troops to their mutual border. Saudi Arabia successfully mediates the conflict.
1978	Egypt, Ethiopia	Ethiopia's proposed construction of dams on the headwaters of the Blue Nile leads Egypt to repeatedly declare the vital importance of water. President Anwar Sadat states 'the only matter that could take Egypt to war again is water'. Egypt's Foreign Minister, Boutros-Ghali states 'The next war in our region will be over the waters of the Nile, not politics.'
1986	North Korea, South Korea	North Korea's announcement of its plans to build the Kumgansan hydro-electric dam on a tributary of the Han River, upstream of Seoul, raises concerns in South Korea that the dam could be used as a tool for ecological destruction or war.
1990	South Africa	Pro-apartheid council cuts off water to the Wesselton township of 50,000 black people, following protests over miserable sanitation and living conditions.
1990	Iraq, Syria, Turkey	The flow of the Euphrates is interrupted for a month as Turkey finishes construction of the Ataturk Dam, part of the Southeast Anatolia Project (GAP). Syria and Iraq protest that Turkey now has a weapon of war. In mid-1990 the Turkish president threatens to restrict water flow to Syria to force it to withdraw support for Kurdish rebels operating in southern Turkey.
1991	Iraq, Kuwait	During the Gulf War, Iraq destroys much of Kuwait's desalination capacity during retreat.
1991	Iraq, Kuwait, USA	Baghdad's modern water supply and sanitation system are intentionally targeted by Allied coalition.
1992	Czechoslovakia, Hungary	Hungary abrogates a 1977 treaty with Czechoslovakia concerning construction of the Gabcikovo/Nagymaros dams project based on environmental concerns principally over reduced volume of river flow. Slovakia continues construction unilaterally, completes the dam, and diverts the Danube into a canal inside Slovak republic. Massive public protest and movement of military to the border ensue; issue taken to the International Court of Justice.
1997	Singapore, Malaysia	Malaysia supplies about half of Singapore's water and threatens to cut off that supply in retribution for criticisms by Singapore of government policy in Malaysia.
1999	Lusaka, Zambia	Terrorist bomb blast destroys the main water pipeline, cutting off water for the city of Lusaka, population 3 million.
1999	Yugoslavia	Yugoslavia refuses to clear war debris on Danube (collapsed bridges) unless financial aid for reconstruction is provided; European countries on Danube fear flooding will result. Diplomats decry environmental blackmail.
1999	Puerto Rico, USA	Protesters blocked water intake to a US Navy Base in opposition to the US military presence and the Navy's use of the Blanco River, following chronic water shortages in neighbouring towns.

Source: Gleick, 2000, pp.187–9.

4.3 Case study: Tigris and Euphrates rivers

In Iraq in 1974, the Euphrates River stopped. Syria had cut off the flow to fill the reservoir behind the new Tabqa Dam. River flow had already been reduced because, at the same time, the reservoir of the new Keban Dam in Turkey was also filling. There was no water left to flow into Iraq. Iraq called up its army and the troops assembled on the Syrian border. In response, Syria hastily released about 200 million m^3 of water into the Euphrates channel and, on that occasion, crisis was averted. This was just one example of the tensions between Turkey, Syria and Iraq, the three main riparian states of the Euphrates and Tigris river basins.

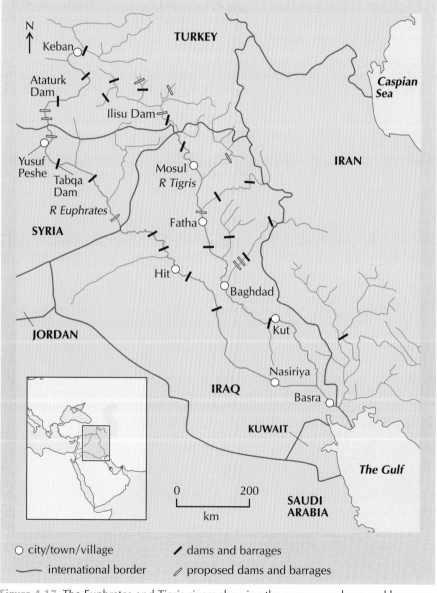

Figure 4.17 The Euphrates and Tigris rivers showing the numerous dams and barrages.

Both rivers rise in the mountains of Turkey and flow southwards. The Tigris also has a few tributaries arising in Iran. The Euphrates, further west, takes a meandering path through Syria and then into Iraq where it meets the Tigris at the confluence, north of Basra. The joint river, now called the Shatt al Arab, reaches the sea at the north end of The Gulf on the border with Iran (Figure 4.17).

Turkey, as the upstream riparian state, has the significant advantage of controlling the sources of both rivers and also many of their tributaries. The climate in the mountains of eastern Turkey is characterized by dry summers and heavy winter precipitation resulting in seasonal extremes in river flow. The people living on the plains in what was Mesopotamia, now part of Iraq, have been subjected to repeated floods throughout history and they have been building dams and barrages to control flooding for thousands of years. In contrast to the Turkish mountains, the climate in the lowlands is much drier. The people living in the plains have been dependent on the two rivers for their water supply, both for direct consumption and for irrigation, for five thousand years or more. The dry climate and lack of alternative water sources are problems that cannot be easily rectified and Syria and Iraq continue to be dependent on the waters of the Euphrates and Tigris.

Nevertheless, Turkey claims its right as the upstream country to own and control the water that originates within its border. Iraq claims historical rights to the rivers because its people have depended on them for centuries. Syria argues on both grounds – claiming ownership rights and historical user rights. Turkey's attitude is demonstrated by the comments of the Turkish Prime Minister at the opening of the Ataturk Dam, one of the largest dams in the region. He said:

> This is a matter of sovereignty. We have every right to anything we want ... Water resources are Turkey's, and oil is theirs. Since we don't tell them 'Look, we have a right to half of your oil,' they cannot lay claim to what's ours ... These crossborder rivers are ours to the very point they cross the border.
>
> (*The Turkish Times*, 15 August 1992, p.5, quoted in Waterbury, 1993, p.57)

The potential for conflict under these circumstances is plain to see. Added to this, the demand for the 'shared' water resource is increasing. The rivers are already well used (Table 4.4) and more projects are in development (Figure 4.17).

Table 4.4   Average yearly flow in the Euphrates and Tigris for the period 1931–1966 (rounded to nearest billion m^3)

River Euphrates	billion m^3	River Tigris	billion m^3
at Keban, Turkey	20	at Mosul	21
at Yusuf Peshe, Syria	30	at Fatha	37
at Hit, Iraq	26	at Baghdad	39
at Nasiriya, Iraq	14	at Kut	31

Source: adapted from Soffer, 1999, p.78.

Activity 4.7

Using the map in Figure 4.17, locate the places mentioned in Table 4.4. Can you explain why the flow increases along the courses of the rivers, and then decreases?

Comment

The map shows that several tributaries join the main rivers in their upper sections and this adds to the total volume. The lower flow downstream indicates how much water has been abstracted from the rivers as they flow through the drier south of the region. The size of the volume reduction makes clear how much water is already being consumed. And note that these data only apply to years up to 1966. Given the trends in increased consumption discussed earlier in this chapter, it is likely that the downstream volumes have reduced further since then.

With cooperation between the three major riparian countries it could be possible to share the resources fairly; this would take account of each country's needs and their alternative water sources. But cooperation is sadly lacking. The lack of trust among the three means they each try to satisfy their water needs within their own borders. The difficulty is simply that there is not enough water in the system to meet all their demands.

Turkey has the most expansive plans of the three. The Southeast Anatolia project (the GAP project) involves the construction of more than twenty dams that will allow irrigation of millions of hectares and produce ample electricity to industrialize the entire region. When the Ataturk Dam was completed in 1990, in a repeat of the 1974 incident, Turkey stopped the flow of the Euphrates for one month to fill the reservoir. Before doing so, as a concession, Turkey had temporarily increased the flow to compensate Syria and Iraq for the forthcoming loss. But the two downstream countries had no means to store the water so the gesture was redundant. In Syria, the people suffered a lack of drinking water, several hydroelectric stations had to be shut down and irrigation was cut back. The Iraqi and Syrian governments insisted that Turkey restore the river flow but Turkey responded by arguing that their reservoir needed to be filled. After a month, Turkey finally did restore the flow and on this occasion the weather helped. Heavy rains and snow during following seasons completed the filling of the reservoir and once again further crisis was averted.

Turkey's vast and grandiose schemes may never reach completion. Several dams are still under construction or are only plans, but if the GAP project is ever completed the reduction in flow would be considerable – up to 40 per cent of the flow in the Euphrates alone would be lost to Syria and Iraq. Not only this, the quality of the remaining water would also be a major problem. Excess water draining from the vast area of planned irrigation in Turkey would flow back to the river and carry salts, fertilizer and pesticide residues with it.

To compound the problem, Syria also has grand plans. More dams to increase electricity generation and to expand the area of irrigated land are priorities for them as well. Iraq is in a similar situation, although information and plans for new schemes are not readily available from Iraq because of its political isolation since the Gulf War of 1991.

The sums just do not add up. The Tigris-Euphrates is already under strain from water demand. If the development plans are followed through then at some point demand will exceed supply. At that point, there will be no water left by the time the river reaches the sea. Estimates of when the crisis point will occur vary, due to uncertainty of existing data and unknown outcomes for plans currently still on the table. Soffer (1999) suggests that by 2030 there will be insufficient water in the system to meet demands (Table 4.5).

Table 4.5   Predicted demand and supply of water in Euphrates and Tigris basins for the years 2000–2040

| | billion m^3 | | | | |
	2000	2010	2020	2030	2040
Demand – Turkey	5.7	12	15	20	27
Demand – Syria	2 to 3	6.2	8.3	10.4	12.5
Demand – Iraq*	46 to 55	46 to 55	46 to 55	46 to 55	46 to 55
Total demand	53.7 to 63.7	64.2 to 73.2	69.3 to 78.3	76.4 to 85.4	85.5 to 94.5
Supply	80	80	80	80	80
Balance (supply minus demand)	+26 to +16	+16 to +7	+11 to +2	+4 to -5	-5 to -14

*Iraqi demands are based on 1980s data and in the absence of updated figures have been assumed to remain constant, although this is unlikely.

Source: adapted from Soffer, 1999, p.106.

○   In Table 4.5, it has been assumed that the total volume of water supply in the river system will remain the same throughout the period shown. Can this be guaranteed?

●   Like other prediction models, it would be foolish to claim 100 per cent certainty that the forecasts are guaranteed. In this instance, there must be some uncertainties associated with climate change that could affect the precipitation in the region and hence the volume of water in the rivers. Whether this could increase or decrease the supply is not known.

Long before the point when the river dries up completely, the flow will be reduced to such an extent that profound ecological damage will be done. A certain minimum flow is essential in all rivers to continue to support the numerous species of plants and animals depending on it. As the volume decreases, the quality decreases too. If the volume of the river falls below approximately 10–20 per cent of its normal discharge, there will be insufficient to drain the wastes from the millions of inhabitants of Baghdad and the plains. The rivers will degenerate into a 'stinking and salty sewage system' (Soffer, 1999, p.105). This dire outcome is not inevitable. Greater cooperation between the three countries, better use of alternative resources and changes to agricultural practice could all play their part. However there are many obstacles, not least that water issues play a political, not just an economic role.

The relationship among the three countries is very intricate – their disputes over water rights and usage are only part of the story and are overlain by complex territorial, historical and cultural differences (see Figure 4.18). Syria and Turkey have border disputes dating back to the Second World War and conflicts over other rivers as well. A major source of tension is the differing attitudes to the Kurdish people, who live in all three countries. Turkey and Iraq oppress the Kurds who fight for independence. However, Syria is home to relatively few Kurds who therefore pose no political threat, and they support the Kurdish rebels. Syria has an added interest in supporting them because one of the Kurdish guerrillas' identified targets is the GAP scheme.

During the crisis sparked by the filling of the Ataturk Dam, Turkey promised to restore the flow of the Euphrates if Syria conceded to certain demands. These were that Syria would stop supporting the Kurdish underground, who were operating from within Syria, and that it would also withdraw support from Armenian groups who were fighting against Turkish authority. Turkey, with the upper hand as far as access to water is concerned, has a very powerful weapon available when it comes to negotiating with its neighbours.

In contrast, the relationship between Turkey and Iraq was quite good until the 1991 Gulf War. They shared opposition to the Kurds and had strong trading relations. However, after Iraq invaded Kuwait, Turkey sided with the coalition against Iraq and provided a base for air attacks on Iraqi targets. Iraq and Syria both oppose Turkish water policies so they have that in common, but there the similarities end. There is a deep ideological conflict between the two and Syria has consistently aligned itself with Iraq's enemies.

With this history, it is hardly surprising that no formal agreement on equitable sharing of their water resources has been made. Given the continuing political tensions and conflicts in the region, it seems unlikely that the situation will improve. Added to this is the uncertainty of climate change. In recent years the overall river flows have been relatively high but if there were one or two dry years, Turkey's consumption is likely to leave Syria and Iraq with insufficient water. This would only exacerbate the existing inequalities of water availability in the three countries. The possibility of a water war is inescapable.

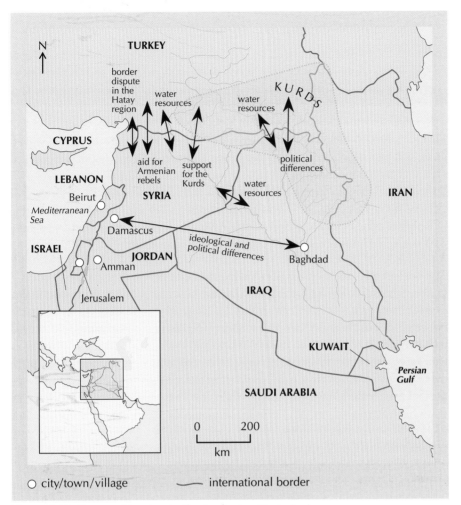

Figure 4.18 Causes of tension in the Euphrates–Tigris Basin.

Summary

Many river basins cover two or more countries and this could give rise to conflict if adjacent nations are competing for use of the water resources. Avoiding disputes requires those countries to agree how to share such rivers by negotiating and cooperating on the principles to be applied. The UN Convention of 1997 established the principle that use should be equitable and reasonable and cause no significant harm to other states.

Disputes over water resources are frequently complicated by other differences between countries. As yet, no war has been fought over water resources but some predict that this will happen in the future. The Tigris–Euphrates basin illustrates many of the complexities of a difficult international water-sharing situation.

5   Conclusion

Although freshwater is renewed by the action of the hydrological cycle, what is certain is that there is always only a limited quantity available and this has to be shared among an increasing number of people for many different purposes. This is at the root of our first chapter question: how and why do contests over water arise? And this question has been our main focus. Multiple uses of a river make for a complex system of differing interests among whom the contests, if they arise, will take place. Competing demands between different users are coupled with the interconnections of the whole water system that link upstream use to downstream impact. Alqueva and Okavango both illustrated all these underlying reasons for conflict and the Tigris and Euphrates story adds another dimension, in that rivers do not follow human territorial borders and this can lead to disputes over ownership, especially where other antagonisms exist.

The second key question on how environmental understandings are produced has been interpreted for this chapter to mean: how can potential water conflict be avoided? One answer is to find a way to integrate the different uses, balance opposing demands and share the resource equitably. Importantly, this should include uses that depend on the ecological and environmental value of water as well as its direct use. River basins need to be considered as a whole system and managed on an integrated, catchment-wide basis. Water resources can only be managed effectively and efficiently when the river basin is considered in its entirety, whether or not this crosses regional or international borders. This requires cooperation between all the various users and interest groups and consideration of the needs of all. The Danube agreement demonstrates how this can be achieved although there are many examples – of which the Turkey, Iraq and Syria relationship is just one – where international collaboration is lacking.

Finally, although it has not featured much here, we should briefly consider the third key question as it relates to this chapter: can inequalities in the availability of water resources be reduced? A slowdown in global population increase would help but is unlikely. Many of the contrasts between regions of the world are unalterable. Dramatic climate change aside, there is very little chance that desert will become floodplain. There are some novel alternative sources of water supply that could possibly alleviate the situation for some. For example, desalination of seawater offers a genuine additional source of 'new' water but consumes large amounts of energy and is very expensive. Other alternative sources, such as towing icebergs from polar regions to areas of high demand, have been proposed but are not generally feasible. A more sustainable approach would be to shift the emphasis from the management of supply, which uses dams, reservoirs and pipelines, to management of demand. Demand management encompasses a variety of techniques for reducing overall water consumption ranging from reducing leakage in pipework to making better use of rainwater. At a domestic level it includes recycling water by using greywater (from baths, basins and washing machines) for toilet flushing, using water-efficient appliances and simple things like cleaning your teeth without leaving the tap running.

In the domain of water resource management, the chance of conflict rises as populations grow and demands increase, while global climate change adds a degree of uncertainty to the situation. These problems are not insurmountable but require the creation of long-term solutions, which can only be achieved by cooperation and agreement between interest groups. As UNESCO's World Water Development Report says:

> The water crisis – be it the number of children dying of disease or polluted rivers – is a crisis of governance and a lack of political will to manage the resource wisely.
>
> Globally, the challenge lies in raising the political will to implement water-related commitments. ... Otherwise water will continue to be an area for political rhetoric and lofty promises instead of sorely needed actions.
>
> (UNESCO, 2003)

References

Abrams Neme, L. (1997) 'The power of a few: bureaucratic decision-making in the Okavango Delta', *Journal of Modern African Studies*, vol.35, no.1, pp.37–51.

Blackmore, C. and Ison, R. (1998) T860 *Environmental Decision Making: A Systems Approach*, Block 2, Milton Keynes, The Open University.

Blowers, A.T. and Smith, S.G. (2003) 'Introducing environmental issues: the environment of an estuary' in Hinchliffe et al. (eds).

Brandon, M.A. and Smith, S.G. (2003) 'Water' in Morris et al. (eds).

DFID (Department for International Development) (2001) *Addressing the Water Crisis*, London, DFID.

ENS (2001a) 'Danube restoration plan takes flight at summit', *Environment News Service*, http://ens-news.com/ens/may2001/2001L-05-01-10.html (accessed March 2002).

ENS (2001b) 'Financial cracks appear in Portuguese dam project', *Environment News Service*, http://ens-news.com/ens/may2001/2001L-05-01-01.html (accessed March 2002).

Furniss, P. and Lane, A. (1990) P588 *Practical Conservation: Water and Wetlands*, Milton Keynes, The Open University.

Gleick, P.H. (2000) *The World's Water 2000–2001*, Washington, D.C., Island Press.

Hinchliffe, S.J. and Belshaw, C.D. (2003) 'Who cares? Values, power and action in environmental contests' in Hinchliffe et al. (eds).

Hinchliffe, S.J., Blowers, A.T. and Freeland, J.R. (eds) (2003) *Understanding Environmental Issues*, Chichester, John Wiley & Sons/The Open University (Book 1 in this series).

Lane, A. et al. (1999) T552 *Systems Thinking and Practice: Diagramming*, Milton Keynes, The Open University.

McCully, P. (1996) *Silenced Rivers: the Ecology and Politics of Large Dams*, London, Zed Books.

Morris, R.M. (2003) 'Conclusion' in Morris et al. (eds).

Morris, R.M., Freeland, J.R., Hinchliffe, S.J. and Smith, S.G. (eds) (2003) *Changing Environments*, Chichester, John Wiley & Sons/The Open University (Book 2 in this series).

Rennie, D. (2002) 'Three Gorges dam a "toxic time bomb" ' *The Daily Telegraph*, 9 March.

Roberts, P. (1995) *Environmentally Sustainable Business: A Local and Regional Perspective*, London, Paul Chapman Publishing, quoted in McCulloch, A. (1996) 'Environmental assessment, the learning society and the search for sustainable development', paper for conference on integrating environmental assessment and socio-economic appraisal in the development process, University of Bradford, 24–25 May.

Savenije, H. and van der Zaag, P. (2000) 'Conceptual framework for the management of shared river basins; with special reference to the SADC and EU', *Water Policy*, no.2, pp.9–45.

Soffer, A. (1999) *Rivers of Fire: the Conflict over Water in the Middle East*, Lanham, Maryland, USA, Rowman and Littlefield Publishers Inc.

Thomas, A., Selolwane, O. and Humphreys, D. (2001) 'The campaign against the Southern Okavango Integrated Water Development Project' in Thomas A., Carr S. and Humphreys D. (eds) *Environmental Policies and NGO Influence*, London, Routledge.

UNESCO (2003) 'Political inertia exacerbates water crisis, says World Water Development Report' UNESCO Press release, http://portal.unesco.org/ev.php?URL_ID=10064&URL_DO=DO_TOPIC&URL_SECTION=201&reload=1047041399 (accessed March 2003).

Waterbury, J. (1993) 'Transboundary water and the challenge of international cooperation in the Middle East' in Rogers P. and Lydon P. (eds.) *Water in the Arab World*, USA, Harvard University Press.

Wilkinson, I. (2002) 'Protests drowned as vast new lake begins to fill up', *The Daily Telegraph*, 9 February.

Chapter Five

Trading with the environment

Annie Taylor

Contents

1   Introduction

One wet morning in December 1999, a crowd of about 50,000 people gathered on the streets of downtown Seattle. It was to become a historic protest, the first of many similar large-scale demonstrations. There were representatives from a huge range of groups: US steelworkers, consumer groups, development groups such as Oxfam, environmentalists such as the Royal Society for the Protection of Birds and Friends of the Earth, and animal rights groups. And it was not just organized groups – there were also individuals with no organized base. The real importance of the protest lies in the fact that so many people came together with a shared and unshakeable belief that the cause of the problems on which they were campaigning relates to economic globalization and more particularly to the management of global trade. They chose Seattle on the first three days of December 1999 (Figure 5.1) because that is where the World Trade Organization (WTO) was holding its Annual Ministerial Meeting.

Figure 5.1 Protestors on the streets of Seattle.

The protestors believed that world trade and the way it was being managed was unfair and harmful to both people and the environment. Of course the other side of this story is that the many hundreds of government ministers, officials and bureaucrats held an equally unshakeable belief that world trade, as it is currently managed, is essentially fair and beneficial to people and the non-human environment. The lines for what is now known as the 'Battle of Seattle' could not be clearer. Moreover you can recognize them in the subsequent demonstrations staged at the major meetings of the WTO and other international economic organizations, for example the IMF Summit in Genoa (July 2001), the simultaneous demonstrations in capital cities around the world coinciding with the WTO Ministerial Meeting in Doha, Qatar (November 2001), and the EU Summit in Barcelona (March 2002).

Global trade is the focus for this chapter and as you read on you will be able to discover for yourself the reasons behind the contests between trade and the environment that were a major part of the protests played out in Seattle. It will also help you to think about whether there is any way that the contests between trade and environment might be resolved.

The links between trade and the environment are not clear-cut. By now you will be familiar with the idea that environmental issues are contested. This chapter aims to introduce you not only to another contested issue but also another level at which environmental contests occur. So far you have covered issues at the local, national and international levels; consideration of trade and the environment will take you to the global level.

The demonstration at Seattle is illustrative of the trade and environment issue and it raises a number of questions:

- Why were so many people demonstrating about environmental and social issues outside a WTO meeting?
- What exactly are the battle lines for the Battle of Seattle? In other words what are the real differences between the views of those inside the meeting of the WTO and those on the outside?
- Exactly how does trade affect the environment in practice?
- If the relationship between trade and environment is contested, whose views prevail and why?

These questions are extremely pertinent to this book and the series of which it is part. The first three questions listed above unpack the first of the book's key questions: how and why do contests over environmental issues arise? The last question listed above offers us a way of developing an answer to the second key question: how are environmental understandings produced? And it provides a springboard to consider the third key question: can environmental inequalities be reduced?

This first section addresses the first in our list of questions. To understand why so many people were demonstrating outside the WTO meeting we need to explore the reasons why the relationship between trade and the environment became an environmental issue. We also need to build up our knowledge of how trade works before we can critically analyse the different viewpoints that were being expressed in Seattle. The second section provides a brief outline of the mechanics of trade. The third section analyses the contests underlying the trade and environment issue in more detail so that we can get a better understanding of the different positions in the Battle of Seattle. The fourth section will apply our understanding of the issue built up in the first three sections to real examples of contests between trade and the environment. This will identify some of the complexities, showing that there is more to the Battle of Seattle than the headline slogan suggests. Finally, the last section will look at the role of power. The issue may be complex but the battle lines are well known and supported by those with different interests and holding different values. The important question then becomes: why does one view prevail?

1.1 Why has the relationship between trade and the environment become an environmental issue?

Figure 5.2 Roses.

Let's look at the example of roses (Figure 5.2). Roses are something we can buy easily – they are sold in our local florists, some restaurants have wandering rose sellers and they can even be bought from the side of the road. They hold a special place in our social lives – something we give as a symbol of our affection. But we rarely, if ever, think about where they came from, how they were grown or how they made the journey to us.

One story tells of Mercedes Posada, one of the many women who work on a rose farm – Agrodex in Colombia. Women are preferred by the plantation managers on the grounds that they are supposedly more nimble-fingered than men and therefore better suited to the tasks of pruning, harvesting, sorting, selecting and packaging (Figure 5.3). The work is hard and hot and involves handling chemical fertilizers and pesticides. These chemicals are some of the most lethal used in agriculture but there are few checks on them because there is no direct risk of them entering the food chain. It is estimated that 20 per cent of the 127 different pesticides used are banned or not registered in the UK or the USA because they are extremely toxic or have been found to cause cancers. The young women who work on these rose farms complain of headaches, nausea and impaired vision and many have suffered miscarriages. Those living nearby are at direct risk from inhaling the chemicals used to fumigate the flowers.

But, it is not just people who suffer: the non-human environment surrounding these rose farms is also badly affected. Forests are cleared to make room for the farms. Once the farms are operating the flowers need lots of water to make them grow and the rose farms draw heavily from local water supplies. Around the town of Madrid in Colombia the water level in the aquifer has fallen from 20 metres to 200 metres, and water now has to come from Bogotá. The pesticides and fertilizers, if not properly treated, run off to pollute local rivers. Because local farmers use the river water for their agriculture, the local produce that they live off – vegetables, grains and grazing pasture livestock – are also affected (Ferrer, 1997; Maharaj and Dorren, 1995).

Figure 5.3 Workers on a rose plantation.
Source: Maharaj and Dorren, 1995.

Activity 5.1

From this example can you think what the negative and positive sides of the trade in roses are? The negative side will involve the negative effects on the workers and the non-human environment. Drawing on your knowledge and understanding of the way human activity can affect the environment, what do you think are the likely effects? The positive side may be related to the income that comes from selling these roses. Again, use your understanding of the previous issues in this book to think about who or what could gain from this money.

There is a useful discussion of uneven development in Castree (2003).

Comment

There are obvious harmful effects from the production of roses. The example clearly demonstrates the potentially harmful effects of pesticides on the health of the workers. In addition the use of pesticides affects the health of those living close to the plantations who risk inhaling the toxic fumes from the chemicals. They are also at risk from contaminated water supplies and from eating the crops and milk produced on land that could be contaminated by the water.

The way roses are grown also has potentially damaging implications for the non-human environment. For example the pesticides used can also affect the animal and plant life that exist beyond the rose plantations. In addition, the loss of forests, cleared for the rose plantations, will have a significant effect on the local ecosystems.

The trade in roses may exacerbate these issues because it sets a distance between the consumer and producer. Consumers are usually largely unaware of the production processes. The chances are that the last time you bought, or were given, a bunch of flowers you were completely unaware of the way in which they were produced. In addition, it is unlikely that you will have known whether or not the production and consumption of those flowers caused any damage to the environment. Because you have no direct knowledge or experience of the negative effects of production, it is unlikely that your own consumption levels will change in response to environmental damage occurring hundreds of thousands of miles away.

Finally, most consumers are untroubled by the levels of pollution generated by transporting flowers (and all other traded items) around the world. This is another 'hidden' effect of trade.

On the positive side, the trade in flowers brings foreign money into Colombia. At a national level, and in spite of the social and environmental effects, flowers are important for Colombia's economy. Flowers are its third most important export crop. Moreover more money can be made from growing and exporting flowers than from most other agricultural crops. At a local level, even though the wages are poor and the working conditions can be dangerous, many workers would face financial hardship if they could not get any work at all.

The example of the trade in roses provides a good illustration of how global trade can affect the environment (see Castellanos, 1997). It shows, too, that trade is not an unquestionable villain. The trade in roses generates economic wealth. It is possible that this money could be used to finance the clean-up of pollution or research into new, environmentally beneficial technologies. The production of roses also creates an income for the workers. If Agrodex increased the wages paid this would give their employees some economic security, leaving the workers freer to devote more attention and resources to caring for the environment. As consumers are awakened to the environmental damage done by certain growing practices, they may press for the use of new technologies that are less environmentally harmful. In this way, global trade can act as the channel through which new environmentally sensitive ideas are transmitted, educating producers and consumers alike.

What we are beginning to see is that the trade–environment relationship is an important one, but that the exact nature of that relationship is debated. Some people emphasize the economic benefits as the major outcome of trade, and play down the environmental damage it can cause. Others, using the same evidence, focus on the negative effects of trade on the environment as outweighing any possible economic gain. Thus, because there is so little unambiguous information available concerning the relationship between trade and the environment, conflicts have arisen between people who hold different ideas based on different values. In order to understand more about how the trade–environment relationship has become a contested issue, we will need to explore these different ideas and values in more detail. First, however, we need to build up a more detailed picture of how trade operates.

Summary

- The issue of trade and environment is a focus for public protests.
- The exact nature of the relationship between trade and the environment is contested.
- The way things are made for export affects both the human and non-human environments.
- These effects may be positive or negative.

2   How does trade work?

Trade is about the exchange of goods and services between individuals or groups. This can take place within local communities, within countries or across state borders. Two economic concepts, now over 200 years old, underpin our understanding of trade:

- specialization of production
- comparative advantage.

Have a look at Box 5.1. You do not need to be fluent with these theories immediately but you might want to refer back to them as your understanding of the trade and environment issue develops.

BOX 5.1 Theorizing trade

Specialization of production

Writing in 1776, the economist Adam Smith argued that things could be produced most efficiently if there was specialization. He believed that by focusing on the production of just one thing, more can be produced and it can be produced more efficiently. However, if production becomes specialized, producers will need to be able to trade in order to sell the extra products they have made and to buy what they have not produced.

Comparative advantage and factor endowments

In 1817, David Ricardo built on Smith's ideas of specialization to develop the theory of *comparative advantage*. Comparative advantage asserts that countries will be better off if they specialize in producing and exporting goods that they are comparatively good at producing and can produce comparatively cheaply, and import goods that they are comparatively less good at producing or produce comparatively more expensively. The outcome, Ricardo argued, would be that all countries would be better off. Ricardo based the costs of production on the costs of labour needed to produce something (the number of 'man hours' it took to produce it). To illustrate his ideas Ricardo used a simple example of the trade in wine and cloth between Portugal and England – two countries trading in two products. In this example, Portugal produces cloth and wine with less labour (and therefore less cost) than England, but it takes Portugal comparatively more labour to produce cloth than to produce wine. Ricardo argued that it is more profitable for Portugal to invest in the production of wine and import the cloth from England.

This may be easier to understand if you think of it in terms of two individuals. Imagine there are two people Ms Wig and Mr Keyin. Ms Wig used to be an excellent typist before she took an OU degree and eventually became a practising barrister; she now earns £48 an hour. Ms Wig uses a good secretary, Mr Keyin, who charges £12 an hour. If Ms Wig needs 10 pages of case notes typed up she could do it for herself and it would take an hour. If Mr Keyin were to do it, it would take two hours. An hour of Ms Wig's time is the equivalent of £48. For this amount of money she could employ Mr Keyin for four hours. Even though he is twice as slow, she would get twice as much typing done for the same amount of money as if she had done it herself.

Although comparative advantage may explain the differences between countries and the reason why countries should trade, it fails to explain why differences between countries occur. *Why* has Colombia got a comparative advantage in roses and not jumbo jets or precision medical equipment? This question was addressed by two economists: Heckscher writing in 1919, and Ohlin who developed Heckscher's ideas in 1933. Taken together, they assert that different countries or regions will be more or less endowed with different factors of production. These they identified as natural resources (land and raw materials), labour supply (skilled and/or unskilled), capital (money needed to buy materials and equipment) and technology. Colombia has a good climate to grow roses; it also has an abundant supply of skilled and semi-skilled manual labour. Therefore, this theory would argue, it makes sense for Colombia to specialize in the production of roses.

Activity 5.2

Now take a moment to consider five things that you have around you: it may be your clothes, your groceries, perhaps even your furniture. Make a note about where these things come from. How many were produced by you? How many were produced locally and how many came from further afield?

Comment

I would guess that the majority have been made, grown or extracted by other people living in other countries. Some of the things may have been manufactured in more than one country and transported across many different countries before they arrived in your home (see Figure 5.4). By stopping to think just how many of the things we have, and rely on, come from abroad we get some idea of how central trade is to our everyday lives.

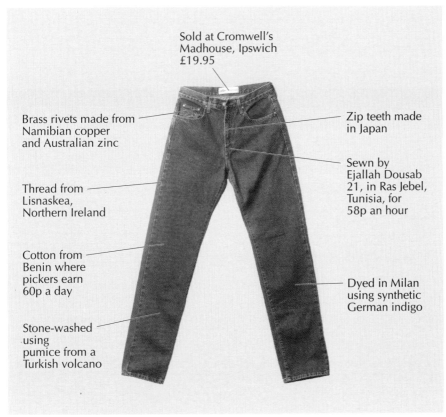

Sold at Cromwell's
Madhouse, Ipswich
£19.95

Brass rivets made from
Namibian copper
and Australian zinc

Zip teeth made
in Japan

Sewn by
Ejallah Dousab
21, in Ras Jebel,
Tunisia, for
58p an hour

Thread from
Lisnaskea,
Northern Ireland

Cotton from
Benin where
pickers earn
60p a day

Dyed in Milan
using synthetic
German indigo

Stone-washed
using
pumice from a
Turkish volcano

Figure 5.4 The global origins of a pair of jeans.
Source: The Guardian (G2), 29 May 2001.

2.1 Globalizing trade

The importance of trade goes beyond our own individual experiences. As the example of the rose farms in Colombia suggests, trade is central to a country's national economy. Indeed, global trade has become more and more important to a growing number of countries. The processes of globalizing trade have become particularly intensified since the 1950s. For example:

- The value of goods traded has increased. Figure 5.5 shows that the value of goods exported increased from US$311 billion in 1950 to US$5.4 trillion in 1998. That is a 17-fold increase in just under 50 years (French, 2000).
- Governments have less control over what is imported and, to some extent, exported. Since the 1980s there has been a concerted drive for greater trade liberalization. Methods to control trade are being withdrawn in order to widen access to markets in nearly all countries around the world. Table 5.1 shows how the number of tariff barriers (taxes on imports or exports) has declined. For example, in the USA it declined from 6.7 per cent in 1965 to 3.5 per cent in 1985 (Held et al., 1999, p.180).

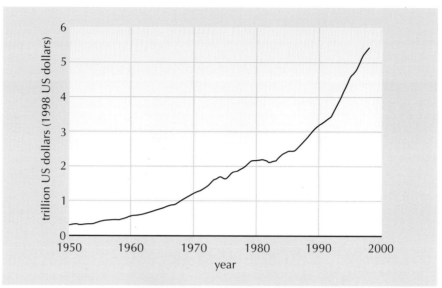

Figure 5.5 World export of goods, 1950–1998.
Source: IMF, in French, 2000, p.8.

Table 5.1   Tariff rates for selected countries, 1965–1985 (percentages)

	1965	1970	1975	1980	1985
France	6.1	2.6	1.4	1.1	0.9
Germany	4.6	3.0	2.4	1.8	1.3
Japan	7.5	7.0	3.0	2.5	2.4
Sweden	6.3	4.0	2.4	1.7	2.5
UK	6.0	2.8	1.8	2.2	1.7
USA	6.7	6.1	4.4	3.1	3.5

Source: EPAC, 1996, p.10, quoted in Held et al., 1999, p.180.?

- Countries are now trading with a greater number of different countries. Figure 5.6 shows that the rich countries of world now have over 120 trading partners and some of the developing countries in South America, Africa and South-east Asia have over 91 trading partners. In fact there is only a small percentage of countries that have less than 30 trading partners. What this suggests is that countries are now firmly connected to a global trading system (Held et al., 1999, p.166).

- The number and power of **transnational companies (TNCs)** have increased in significance. The United Nations Development Programme reported that TNCs are involved in 60 per cent of global trade. One third of this trade is intra-firm trade, i.e. trade *within* transnational companies. (United Nations Development Programme, 1999). A report from the United Nations Research Institute for Social Development in 1995 stated that TNCs controlled over 75 per cent of all world trade in commodities, manufactured goods and services (United Nations Research Institute for Social Development, 1995, p.27).

A **transnational company** (TNC) is a profit-making company that controls assets and conducts business in a number of different countries.

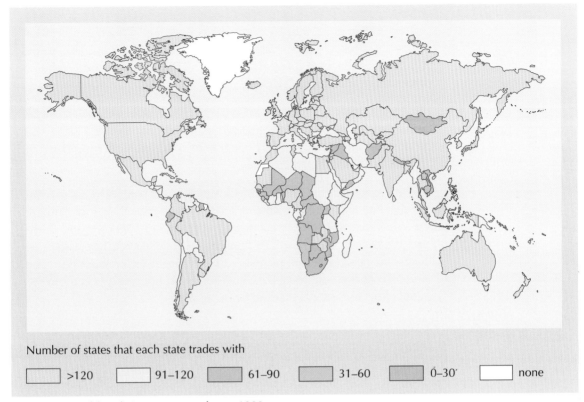

Number of states that each state trades with

| >120 | 91–120 | 61–90 | 31–60 | 0–30' | none |

Figure 5.6 World trade interconnectedness, 1990.
Source: Held et al., 1999.

In practice, these changes have meant that goods and services produced in one country are in direct competition with those produced in another country. To give one small example, I have often wondered why the supermarkets near me will often only sell potatoes from Italy and none from the nearby farms. Perhaps you have noticed similar anomalies. It seems to suggest that we are now consumers (in my case, of potatoes) within a global marketplace (in my case, a 'global market for potatoes'), mainly under the auspices of the WTO (see Box 5.2 and Figure 5.7).

BOX 5.2 The WTO and TRIPS and TRIMS

Since the end of the Second World War there has been a mechanism for agreeing and implementing regulations to oversee world trade. From 1948 to 1995 this was done under the auspices of the General Agreement on Tariffs and Trade (GATT). Since 1995 these duties have been passed on to the World Trade Organization. Like its predecessor GATT, the WTO is guided by liberal economic values that prioritize economic growth and increasing **trade liberalization**. Trade liberalization is the reduction in barriers to trade. Governments agree to lower or eliminate controls on what products can be imported into and exported from their country and how much can be imported or exported.

trade liberalization

However, the WTO differs from Gatt in that it has much greater powers to enforce compliance with its regulations. Also the WTO has a broader scope to cover not only traditional goods and services but also **trade related intellectual property rights (TRIPS)** and **trade related investment measures (TRIMS)**.

TRIPS are rights of ownership such as copyrights and patents that can be applied to goods or services that are exported. The aim of TRIPS is to protect the intellectual property (such as recorded music or inventions) of private individuals and companies.

TRIMS are measures used by host-country governments to control the amount and nature of investments made by foreign nationals, by limiting the conditions and restrictions that governments can put on them. There are two broad categories. The first covers incentives to invest such as loans, tax rebates or the provision of services at favourable rates. The second covers requirements that investors must fulfil such as limits to the number of components that can be imported.

Figure 5.7 The headquarters of the World Trade Organization, Geneva.

2.2 How the WTO operates

Figure 5.8 shows the organizational structure of the WTO. The WTO is made up of 144 member countries (as at January 2002) that meet to negotiate trade agreements. The Ministerial Meetings, such as the one held in Seattle, are part of this process. When an agreement is reached the member countries will sign the

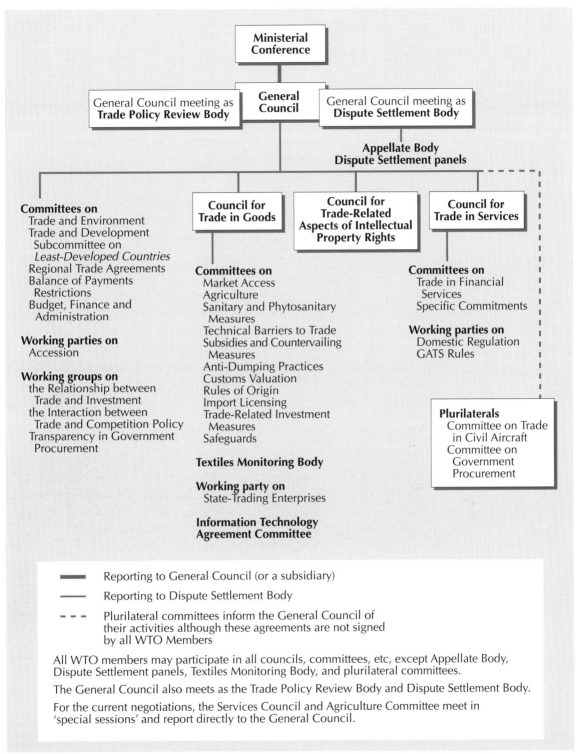

Figure 5.8 WTO organizational structure.
Source: WTO.

agreement and then take it back to their own countries to be ratified and passed into law. In effect the rules of the WTO, reached by negotiation, form a kind of legally binding contract.

Any input from organizations and individuals is expected to come via their own governments. However, as environmental and other non-governmental organizations (NGOs) have increased their interest in the activities of the WTO, the WTO has started to hold regular meetings with NGOs that lie outside the formal negotiations between states.

Note that the WTO defines NGOs as any organization that is not to do with government, thus lumping environmental groups in with TNCs. This is different from the understanding used elsewhere in this series.

The WTO and the environment

In 1992, GATT published a report on trade and environment that declared them to be 'mutually supportive'. In 1994, the agreement that created the WTO declared its commitment to the 'objective of sustainable development'. In 1999 the WTO's Special Study on *Trade and Environment* (World Trade Organization, 1999a) refined this view a little. It accepts that trade and environment are mutually supportive but takes this idea on to develop what they call a 'win–win' solution that favours a joint approach using multilateral agreements (agreements reached between the governments of three or more countries) to protect the environment, combined with the elimination of trade-related measures that distort free trade and have a detrimental effect on the environment.

Summary

- Trade is important at all levels of our lives – local, national and global.
- We are still using ideas from two hundred years ago to understand how trade works – *specialization*, which explains the need to specialize at a local, national and global level, and *comparative advantage* together with *factor endowment*, which explain why we specialize in the way we do.
- Since the 1950s trade has become increasingly global.
- The World Trade Organization is the only worldwide organization set up to manage the rules of trade between countries.

3   What are the contests between global trade and the environment?

So far we have built up a clear idea that the relationship between trade and environment is contested. We also have an understanding of how global trade operates. This section will use this knowledge and understanding to examine the contests between the different views in more detail. From this we will be able to further our understanding of the contests that were played out in Seattle and were enacted at subsequent demonstrations.

As we analyse the conflicting views of the relationship between trade and environment it will be important to identify two key points on which they differ. So far we know it is a contest between those who believe in the benefits of liberalized trade and those who express concerns about the environment. This is mostly about their preferred outcomes. But these preferences are based on their particular view of the world and how it operates. To find out more about this we need to know whom they view as the main actors and how they see power and the operation of power. (In the context of trade this relates to the power to allocate resources.)

Activity 5.3

Take a look at the following matrix. This sets out a grid for organizing our thoughts as we go through this section. As you progress, keep returning to this page and fill in the blank spaces.

	LIBERAL	RADICAL	REFORMIST
Main actors			
Power			
Process			
Preferred outcomes			
View of trade and the environment			

3.1 View 1: Those who emphasize greater trade liberalization – a liberal economic understanding of trade and environment

Liberal economics is important as a *theory* because it is used often and widely to explain how trade operates. It is also important in a more *practical* way because it is used to guide policy formation and to justify trade policies (Figure 5.9). The liberal economic view of trade is guided by three central principles.

Figure 5.9 Trade promotion in West Africa.
Source: The Ecologist, 1999. p.155.

1   Wherever possible, the allocation of resources is best left to the market. This view is based on an idea from Adam Smith (see Box 5.1) who argued that the economy can operate smoothly without any apparent control – only the 'invisible hand' of the market.

2   The market can only function properly where there are clear private property rights. If something is clearly 'owned' it can be traded and the market can regulate its use. However, goods and services that have no clear ownership cannot be so easily traded and their use is difficult to regulate. (This relates to the notion of the 'tragedy of the commons'; see **Castree, 2003**.)

3   Economic growth is essential for human welfare. Liberal economics sees growth as gradual and continuous; the long-term trend will always be increased growth. At a global level, liberalized trade under free market conditions is seen as an important engine behind economic growth. However, the liberal economic position accepts that not all countries and not all people within a country will experience the gains equally. Instead they argue that all people will be better off overall (they talk of *absolute gains*), with some being relatively more better off than others (these are *relative gains*).

This offers an outline of the liberal economic understanding of the economy and of trade. However, in order to further our understanding of the contests over the relationship between trade and the environment we need to ask how the environment and, in particular, the relationship between the environment and trade are conceptualized within such an outlook?

A liberal economic understanding of trade and environment

The first thing to remember is that liberal economics dominates our understanding of how trade operates and also guides policy makers in defining trade policy. When concerns that global trade might be harming the human and non-human environment came to the fore at the end of the 1980s and beginning of the 1990s there was a swift reaction from some policy makers and leading researchers. Liberal economic theory on trade was clarified to explain how concerns about the environment could be addressed without significantly altering the basic liberal economic ideas. One of the first reports addressing the trade and environment issue was the GATT *Report on Trade and Environment* (1992). This claimed that if there were any links between trade and environment they were indirect. It also warned that including environmental issues in the management of trade would risk opening the door to trade protectionism and threaten the health of the global economy. Finally, it asserted that trade increases wealth and opens up channels of communication which, in turn, would benefit the environment. In 1999 the WTO published its Special Study *Trade and Environment* (World Trade Organization, 1999a) that claimed a compromise approach for 'win–win' outcomes that would benefit both the environment and trade.

Activity 5.4

Have a look at the following excerpt from a press release about the WTO's Special Study *Trade and Environment*, which outlines some of the main points. Think about how each of the bullet points relates to liberal economic theory concerning trade and the environment, and make notes in the matrix in Activity 5.3.

- Most environmental problems result from polluting production processes, certain kinds of consumption, and the disposal of waste products – trade as such is rarely the root cause of environmental degradation, except for the pollution associated with transportation of goods;

- Environmental degradation occurs because producers and consumers are not always required to pay for the costs of their actions;

- Environmental degradation is sometimes accentuated by policy failures, including subsidies to polluting and resource-degrading activities – such as subsidies to agriculture, fishing and energy;

- Trade would unambiguously raise welfare if proper environmental policies were in place;

- Trade barriers generally make for poor environmental policy;

- A good environmental profile is often more of an asset for a firm than a liability in the international market-place, notwithstanding somewhat higher production costs;

- Economic growth, driven by trade, may be part of the solution to environmental degradation, but it is not sufficient by itself to improve environmental quality – higher incomes must be translated into higher environmental standards.

(World Trade Organization, 1999b)

Comment

It is clear that this report upholds the fundamental principles of a free market supported by private property rights and economic growth. The underlying message is to counter any suggestions that there is a direct relationship between trade and the environment.

In particular, the first three points clearly absolve trade of any direct link with environmental degradation. The environmental degradation is the result of production and consumption patterns and trade is seen as a neutral channel for exchange.

The fourth point highlights the importance of trade because it increases human welfare – the human environment. A liberal economic perspective is fearful that any trade barriers could, potentially, threaten this and the fifth point hints at this fear.

The sixth point reiterates the potential benefits that trade can bring to the non-human environment by increasing wealth and encouraging production methods that are more sensitive to the non-human environment.

The final point repeats the general view that trade can only have an indirect effect on the environment and changes the focus to better and more effective international environmental regulations.

3.2 View 2: Those who prioritize the interests of the human and non-human environment – the environmental perspective

Even though the liberal economic perspective is widely accepted, the demonstrators outside the WTO meeting in Seattle established in people's minds that there are alternative views. Many of those demonstrators were involved in conceptualizing an alternative to the current system of global trade that they felt could take better account of environmental and social issues. This work is important for two reasons: first and foremost because it presented a critique of the existing system, and second because it offered alternative ways of organizing global trade (although these ideas had only ever been applied on a small scale, if at all).

However, there is a great diversity within the environmental perspective. I find it helpful to classify those who present an environmental focus into two broad groups: the radicals and the reformists. **Radical environmentalists**, sometimes called rejectionists (Williams, 2001, p.55), are groups and individuals who reject the existing global economic and political order. They are reluctant to engage in discussions with the existing institutions and, instead, call for their replacement. **Reformist environmentalists**, sometimes called accommodationists (Williams, 2001, p.55), are those who adopt a compromise position in the trade and environment debate. They engage in discussions with existing institutions and try to work towards reform of the existing global economic and political structures.

radical environmentalist

reformist environmentalist

View 2a: A radical environmentalist approach

In this category there are many different groups. Some of the more active in the trade and environment debate include the radical US-based groups Public Citizen and the International Institute for Agricultural Policy. There are also some more widely known organizations such as Greenpeace and a host of smaller groups that emerge in response to local situations.

Activity 5.5

Have a look at the following excerpt from a 'Sign-On Letter' posted on the Public Citizen's web page in 2001.

While you are reading it see if you can identify aspects of the liberal economic perspective that it is criticizing. It may help to refer back to your matrix: think about the headings in the left-hand column and anything you may have filled in for the liberal economists.

> It's time to turn trade around. In November 1999, the World Trade Organization's (WTO) Third Ministerial Meeting in Seattle collapsed in spectacular fashion, in the face of unprecedented protest from people and governments around the world. We believe it is essential to use this moment as an opportunity to change course and develop an alternative, humane, democratically accountable and sustainable system of commerce that benefits all. This process entails rolling back the power and authority of the WTO.
>
> The GATT Uruguay Round Agreements and the establishment of the WTO were proclaimed as a means of enhancing the creation of global wealth and prosperity and promoting the well-being of all people in all member states. In reality, however, the WTO has contributed to the concentration of wealth in the hands of the rich few; increasing poverty for the majority of the world's peoples, especially in third world countries; and unsustainable patterns of production and consumption.

The WTO and GATT Uruguay Round Agreements have functioned principally to pry open markets for the benefit of transnational corporations at the expense of national and local economies; workers, farmers, indigenous peoples, women and other social groups; health and safety; the environment; and animal welfare. In addition, the WTO system, rules and procedures are undemocratic, un-transparent and non-accountable and have operated to marginalize the majority of the world's people.

All this has taken place in the context of increasing global instability, the collapse of national economies, growing inequity both between and within nations and increasing environmental and social degradation, as a result of the acceleration of the process of corporate globalization.

The governments which dominate the WTO, especially the United States, the European Union, Japan and Canada, and the transnational corporations which have benefitted from the WTO system have refused to recognize and address these problems. They are still intent on further liberalization, including through the expansion of the WTO, promoting free trade as a goal in itself. In reality, however, free trade is anything but 'free'.

The time has come to acknowledge the crises of the international trading system and its main administering institution, the WTO. We need to replace this old, unfair and oppressive trade system with a new, socially just and sustainable trading framework for the 21st Century.

We need to protect cultural, biological, economic and social diversity; introduce progressive policies to prioritise local economies and trade; secure internationally recognized economic, cultural, social and labor rights; and reclaim the sovereignty of peoples and national and sub-national democratic decision-making processes. In order to do this, we need new rules based on the principles of democratic control of resources, ecological sustainability, equity, cooperation and precaution.

(Public Citizen, 2001)

Comment

I have identified two areas of liberal economics of which the radical environmentalists are critical.

First, I got a strong sense that these radical environmentalists are very concerned about the power relations within trading relationships. In particular they are concerned that the developed countries have the power to shape trading relations to serve their own interests regardless of the interests of the poorer and less powerful. Moreover, they identify the influential and powerful not just as countries but also as other social organizations – most of all TNCs. They identify the less powerful not just as weaker countries (the developing countries) but also as the poor *within* countries and non-human actors, most notably the environment.

Related to this, I noticed that they are concerned with the power to control the agenda, in particular the power to promote a liberal economic agenda for further trade liberalization. We will examine how the role of power is understood in more detail in Section 4.

Second, I saw a more specific dissatisfaction with the priority given to market forces within current global trading relations, in particular a criticism of the often-stated aim to foster a free market within global trade. In particular the letter rejects the claims that a global trading regime based on liberalized markets will ultimately bring wealth to all peoples. In addition, the letter highlights continuing distortions in the market, such as import barriers and government subsidies that have a negative effect on the environment.

Aside from the criticisms of the existing system of global trade, this letter also briefly summarizes its priorities for an alternative trading system. This would replace the market as the driving force with other social and environmental priorities. Most importantly, it appears likely that such a vision could not be accommodated into the existing system of trade through piecemeal reforms. The agenda is therefore for radical reform or complete replacement of the existing system of global trade. As the title of the letter states – their sights are set to 'shrink or sink' the WTO.

View 2b: A reformist environmentalist approach

Like the radical environmentalist approach, the reformist approach also covers a broad spectrum of views: from environmental NGOs such as Friends of the Earth, who are quite critical of the current system of global trade, to the World Wide Fund for Nature (WWF) and environmental economists and international environmental lawyers, who tend to be more accommodating. What distinguishes them from the radical environmentalists is their problem-solving approach. A problem-solving perspective identifies immediate problems without considering how the wider political, social and economic structures might contribute to the problem. It tries to find a solution to the problems identified without trying to alter the existing structures. This means that instead of seeking to change the structure of global trade, they are looking to find compromises that would ameliorate any problems in the relationship between trade and the environment. They believe that it is possible to reconcile environmental concerns with the demands for economic growth.

Activity 5.6

Look at the WWF press report below. It was published on the first day of the WTO Ministerial Meeting in Seattle and summarizes a speech given at the WTO Symposium on International Trade Issues held in Seattle. Using the matrix in Activity 5.3, think about the following questions:

- How does this press report explain the trade and environment issue?
- What methods does it favour to deal with the conflict of interests between trade and the environment? And what does it hope to achieve?
- How would WWF define power?

WTO Symposium on International Trade Issues 2000

Claude Martin, Director General of WWF spoke on a panel at the WTO Seattle Symposium on International Trade Issues in the First Decades of the Next Century. Addressing a large number of government officials, NGOs and journalists present for the session on 'Evolving public concerns and the multilateral trading system', Dr. Martin emphasized that WWF's objective in the context of the WTO is to make trade ecologically sustainable.

Key elements from Dr. Martin's speech include:

- 'We are convinced that multilateral trade rules are a necessity for sustainable development, but those rules have to engender trust.'

- 'WWF believes that the WTO has begun to become both more transparent and more environmentally sensitive over the past few years.'

- 'The new WTO report on Trade and Environment was a positive development, recognizing important realities. It identifies the importance of international cooperation, and multilateral environmental agreements (MEAs), to protect the environment from global and transboundary threats.'

- 'If the WTO does not take concrete steps to initiate environmental reform in Seattle, it will become impossible for any in the environmental community to feel anything but concern about the trade negotiations which flow from this meeting.'

- 'Elimination of agricultural export subsidies, and reduction of other production subsidies would help to reduce overproduction and associated environmental damage in many developed countries.'

- 'WTO members will have to move on transparency generally. Both to regain public confidence and to ensure that they have the informed, democratic consent of their citizens for future trade agreements.'

- 'WWF believes that it is vital to seek more innovative ways to integrate environment and development objectives within the WTO.'

In conclusion, Dr. Martin said: 'WWF does not pretend to have all the answers to the challenges of integrating trade, environment and development policies, nor to be able to find them without dialogue and cooperation with all the people in this room. What we can promise you is that we will be consistent in our protection of the environment. We will oppose green protectionism, just as we will continue to work for the reforms of WTO rules necessary to prevent them overriding legitimate and necessary environmental policies, at both the national and international level. WWF does understand that we will have to find that balance, and rapidly develop new multilateral rules that promote sustainable trade. We hope that all WTO members understand that too – for the sake of the environment and the institution.'

The Director General of the WTO, Mike Moore spoke at the opening of the Symposium. In his speech he specifically mentioned WWF

'Those who oppose and protest are not all bad or mad! Many want to improve the WTO. Others want to capture it to reflect their interests – which is a form of flattery I suppose. Most seek honest engagement. The World Wide Fund for Nature – to take just one example – has made several constructive suggestions about improving the interface between trade and the environment. We should listen, reflect, then act.'

(WWF, 1999)

Comment

The approach of WWF is one of compromise. It seeks to accommodate and increase its involvement in the processes of the WTO. (The response of the then Director General of the WTO welcoming the input from WWF appears to confirm this.)

As I was reading this press report I highlighted five key points:

- The report is clearly accepting of the underlying economic structures.
- There is a clear message that environmental problems need to be resolved through international negotiations and **multilateral environmental agreements (MEAs)**.
- I noticed a focus on redirecting global trade to encourage those rules and practices that had a good effect on the environment.
- WWF are pushing for more participation from non-state actors, in particular the involvement of individuals ('citizens'), groups and associations that have a stake in the work of the WTO.
- They raise the possibility of introducing trade measures to *protect* the environment as a last resort measure.

> Multilateral environmental agreement (MEA) is an agreement concerning environmental issues, reached by three or more countries.

The three perspectives – liberal economic, radical, reformist – which we have now briefly reviewed have been central to the debate over the relationship between trade and the environment. As such, they provide the backdrop to the protests at Seattle and the similar demonstrations that followed. The matrix that you have been using to identify key differences also provides you with a useful summary. To check the information you have, take a look at the completed one at the end of this section. Above all, it shows that each of the three perspectives is based on a different understanding of what is at stake and what interests are supported or promoted.

Completing the matrix has given us a clear reference for the points of conflict. However, at the moment it is quite a 'dry' piece of information. We need to put it to the test by relating it to real examples of where the priorities of trade and environment clash.

Summary

The completed matrix below provides a summary for the section.

	LIBERAL	RADICAL	REFORMIST
Main actors	Governments and governments working within international organizations	TNCs and other large private economic actors e.g. banks; communities, trade unions etc. considered important but relatively powerless	Governments, International organizations, NGOs
Power	Hidden hand of the market	Economic interests	Market plus unequal political and economic capabilities
Process	Management of resources through market forces	Challenging the existing global economic order (including the management of trade)	Negotiation and bargaining about the management of resources
Preferred outcomes	Status quo	New world order	Intervention for better management of the allocation of resources
View of trade and the environment	Trade benefits all through generation of wealth; trade has no direct effect on the environment	Trade, as it is currently managed, perpetuates uneven development and is directly damaging to the environment	Trade benefits all but it can damage the environment if there is no regulation to protect the environment

4   How are the contests between global trade and the environment played out?

Let us now use these three different perspectives to further our understanding of the complex interrelationship between global trade and the human and non-human environment.

4.1 The global trade in roses

To refresh your memory, take another look at our discussion of the rose trade in Section 1.1, which highlights the harmful effects of the current methods of growing roses for export on both the human and non-human environment. But it also identifies some important positive effects for the national economy of Colombia. One thing it illustrated was that the relationship between global trade and the environment is something that is open to interpretation. Now we know a little more about the different perspectives it gives us a chance to reflect on this again.

Activity 5.7

Imagine you have a liberal economist (LE), a radical environmentalist (Rad) and a reformist environmentalist (Ref) in your room. If they were to discuss the global trade in roses, what do you think they would say? How would they view the trade? How would they view the effects on the human and non-human environment? What would they disagree about?

Comment

The discussion may have taken the following form.

The first person to speak is the one with a liberal economic perspective, who starts by explaining the benefits of the global rose trade.

LE   It generates wealth for the company, country and workers and gives access to resources for importing countries and consumers. The Colombian government and, to some extent, the Colombian people, the companies growing roses and the workers all benefit from the money made by the export of roses.

To support this view the liberal economist explains:

LE   Colombia's involvement in the global rose trade is simply a consequence of comparative advantage and factor endowment. The production costs of growing roses for export are relatively low in Colombia because the factors needed are cheap and readily available. For example, there is a good, cheap unskilled and semi-skilled labour supply. Also, the soil and climate in the rose growing area makes it possible to grow roses quickly and relatively easily. Moreover, the pesticides and herbicides needed to bring the blooms up to export standard are easy to get hold of and can be used almost at will because of the low level of regulation.

Rad and Ref   [Both start to mutter about environmental damage]

But the liberal economist goes on to explain:

LE   The liberal economic perspective is not blind to the environmental damage attributed to the rose plantations. However, such damage is the result of producers and consumers not paying the full costs of production and in particular, the costs of minimizing the pollution inflicted on the surrounding area and the costs of cleaning up the pollution. Producers and consumers do not pay the full costs because no one is asking them to. This is partly because there are no clearly defined property rights surrounding air and water. Of course the best solution would be for the Colombian government to introduce taxes and regulations to correct the problem.

Rad   But the Colombian government could not do this because that would lose their 'comparative advantage' of weak regulations concerning the use of chemicals.

LE   That may or may not be true in this case but evidence from the WTO suggests that companies rarely relocate just because of environmental regulations [1]. Even if it were true, the fault is not with global trade, it is because of the lack of multilateral agreement about how to tackle such cases of pollution. If all the governments in a similar position could agree and implement standards for pesticide use then the problem would be solved.

[1 World Trade Organization, 1999a]

Rad   But what about the workers? You've not said anything about them.

LE   The reason I have not addressed the issue surrounding the plantation workers is that, like the pollution of the natural environment, it is not the direct result of global trade. The working conditions within countries have never been of any concern to those who trade with that country. Global trade is about the exchange of goods across national boundaries; it does not *cause* low wages or poor working conditions. These are matters for domestic politics.

Rad and Ref   What about the solutions, then?

LE   I have already offered some solutions that place responsibility on the government of Colombia and on multilateral negotiations. The most important principle in any search for solutions is that they do not become a form of green protectionism (import and export barriers introduced by one country against others on the grounds that they will protect the environment). This would be counterproductive – damaging trade and leaving everyone worse off.

But what about you two, how would you deal with the situation? How would you keep the wealth gained from global trade while solving the unfortunate side effects on the environment?

The radical environmentalist is the next to speak.

Rad   Of course, you could see Colombia's export of roses to be an economic success story. Newspaper reports quote the export earnings to be $600 million a year [2]. It is now the third most important export crop after coffee and coca. The government policy is to support non-traditional exports and the flower growers have been encouraged by cheap credit and exemption from tariffs on goods (such as pesticides) that they have to import [3].

[2 Watkins, 2001]

[3 Maharaj and Dorren, 1995]

On the other hand, this is not the only story. It is the liberalization of global markets that has enabled companies to go into countries and exploit their land and climate for profit. They do not pay for the damage to the human environment – the social and economic costs of ill health – and they do not pay for the damage done to the non-human environment – the polluted rivers and land, the contaminated crops and livestock, and the loss of forests. The market does not account for environmental resources and it is clear that in this case the economic tools to remedy this have not come into play. Global competition, and fear of that competition, means that governments are reluctant to impose environmental standards in case these will put up the costs of production and effectively price them out of the market.

The economic benefits of the global trade in roses go to the large companies that grow the roses. The low costs of production mean they can reap higher profits. The domestic economic policies of the Colombia government support the economic interests of these companies by providing cheap credit and reduced tariff levels. Moreover, most of these companies are not Colombian; they are large, foreign-owned transnational companies. So, in reality, the money does not go to the Colombian national economy but is absorbed into the TNCs. The rules of the WTO also support the economic interests of these companies by ensuring there are few impediments to their export trade.

The economic and social benefits of access to a ready supply of cheap cut flowers go to the consumers in the rich developed countries. But cut flowers are seen as a luxury product so these consumers tend to be the wealthy people in the importing countries.

The environmental side effects from growing roses are not felt by the consumers, or by the majority of the senior personnel in the rose growing companies. They fall on the workers and the poorer populations of Colombia who need the work (however badly paid) or who cannot afford to move away from the pollution. The reality for the workers and other individuals who live and work near the rose growing area is of deteriorating health, higher pollution levels (of water and air) and decreasing water supply. Local laws to protect workers are usually flouted by the large rose growers [4]. [4 Ferrer, 1997]

The immediate problem needs to be addressed by consumer boycotts and sanctions against countries that violate pesticide standards and workers' rights. But the long term also needs to be addressed. We need to develop a different way of trading that supports a different system of production and consumption. The solution is to be found in organizing economic activity so that it promotes and prioritizes local economies and local trade.

At this point the liberal economist interrupts.

LE Such moves are very like the policies that people in the rich, developed, countries of North America and Europe would like to introduce in order to protect their own industries against cheaper imports coming from developing countries. Restrictions on imports or exports to protect the environment are really another form of trade protectionism. You have laudable objectives but how will you stop people applying restrictions on imports and exports, saying they are to protect the environment when the real reason is to protect their own industries.

Then the reformist environmentalist steps forward to speak.

Ref   It is clear that the two views expressed are far apart – they are like ships passing in the night, there does not seem to be any common ground.

The best way forward is to develop an approach that recognizes the different interests of the people directly affected - in other words, the stakeholders. This means the workers, the local populations and those NGOs who speak for the environment, but also the rose growing companies, the governments of both the exporting and importing countries, and the consumers.

With this as a starting point, the priority will be to find a compromise solution. Attempts to stop or restrict that trade will harm the interests of the companies, the exporting governments and ultimately the workers, who will lose their jobs if the trade falls off.

A step away from trade restrictions can be in everybody's interest. Take one example: the USA has changed its trade policy toward South America and dropped import restrictions on cut flowers and other goods as part of a drive to discourage the trade in illicit drugs and this has had some beneficial effects in this direction.

The reformist goes on to suggest possible ways forward.

Ref   Companies could be encouraged to agree to codes of practice that limit the use of chemicals and to implement better working practices such as fairer pay, holiday entitlement and access to medical care. The consumers and the governments of importing countries, where better working conditions already exist, could press for these changes. The 'buying power' of consumers could be harnessed and the governments of importing countries could use multilateral organizations such as the WTO or the United Nations Environment Programme to put political pressure on the Colombian government to enforce these codes of practice.

To support these ideas, the reformist tells of examples where this has happened.

Ref   NGOs within the importing countries of Switzerland and Germany have been instrumental in getting Ciba Geigy and AgrEvo to withdraw several products listed by the World Health Organization as 'highly toxic'. In Colombia, the Colombian Association of Flower

Exporters has drawn up a voluntary code of conduct covering workers rights and pesticide use.

At this point the radical environmentalist interjects.

Rad   That sounds all very good, as if some progress has been made, but there is little evidence that anything has changed. The trade in roses is still damaging the people and the natural environment of Colombia. The Colombian government is too weak to implement any such agreements and while the rose growing companies can use cheap practices they will, because their interests lie in maximizing their profits. Your approach will never succeed because the problem is caused by the current economic structures and you have said nothing about those.

Ref   I agree that the changes have not had an enormous effect and the economic structures do need to be changed but you cannot overhaul the entire system – it is not feasible and may not be desirable. After all global trade has brought greater wealth and choice to many people all around the world.

It is better to reform the system without destroying it completely. The best approach is by multilateral negotiations and agreements. The WTO now has almost 200 members, very nearly every trading country in the world. It sets rules for the management of international trade by negotiation and agreement between the members. Instead of scrapping this it would be better to change the rules of the WTO, for example by ensuring that WTO rules supported arrangements that enabled roses to be grown with less damage to the environment. Of course, this will be more likely if the WTO becomes more transparent and accountable and open to dialogue with NGOs. And this must be our priority.

So what has this discussion told us about the relationship between environmental damage and the global trade in roses? Clearly it is quite complex. However, by looking more closely at some key questions we can build up a clearer picture of the complexities involved.

Is there a relationship between global trade and the environment?

Even this is contested. The liberal economic view plays down the links between trade and the environment arguing that the relationship is, at best, indirect. Radical and reformist environmentalists assume that global trade is a contributory factor to environmental degradation, although they concede that it may not be the only one.

Can trade ever benefit the environment?

All three perspectives would argue that trade may be beneficial to the environment but for quite different reasons. The liberal economists emphasize the increase in wealth that is a result of trade. This, they believe, leads to increased awareness of environmental issues and a greater capacity to avoid damaging the environment or to 'clean up' any damage caused. It also facilitates the transfer of technology needed for cleaner production processes. Reformists would also highlight the need for trade to increase wealth for greater environmental protection and the role of trade as a means to transfer knowledge for example, via consumer actions. Radicals argue that trade in its current form is socially and environmentally harmful. However, they also believe that trade can bring many benefits if it is managed in a way that emphasizes social and environmental justice.

Is trade the main cause of environmental damage in the Colombian rose growing area?

The liberal economic view cites other factors as being more important, for example the lack of property rights to protect the environment, the failure to ensure that the cost of roses includes the cost to the environment of producing them, and the government's failure to enforce labour laws. The reformists do not apportion blame in quite the same way but nevertheless their solutions are similar to the liberal economists: codes of practice for industry, backed by government action to enforce them. The radical view sees trade as the embodiment of an unequal economic system that puts the economic interests of powerful actors above the interests of the environment, the workers or the local populations.

How can the Colombian government reconcile their need for development with the trade and environment issue?

This question is especially interesting because it brings in a perspective that we have not directly addressed so far. The Colombian government and the population of Colombia are faced with very real development concerns. For example not everyone in Colombia has enough to eat, somewhere to live and access to necessary medical care. Faced with these pressing needs, the liberal economic, reformists and radical environmentalists debate on trade and environment may seem of little relevance. Moreover, the government and people of Colombia may feel that the trade and environment issue is being discussed in a way that reflects the values and concerns of developed countries and will therefore serve the interests of those countries. In reality the government and people of Colombia are faced with stark, short-term choices between wealth generated through trade and protection of the human and non-human environment. As far as the government is concerned it may seem more important to reach a satisfactory stage of economic development before environmental issues can be prioritized. Moreover, it may not be possible to follow a more environmentally sustainable development path without outside financial and technological assistance. The workers and local population associated with the rose growing area may wish to see reforms that would give them greater control over their natural resources rather than being at the behest of large foreign-based TNCs. Ultimately both the government and the people feel they are battling against unequal power structures. But we will consider more of this in the next section.

Activity 5.8

Using this analysis of the global trade in roses, how can it help you to understand some of the other contests that centre on trade? Section 4.2 provides an outline of another area where there are conflicts about the relationship between global trade and the environment: the global trade in grain. As you read it think about how those associated with the different perspectives might understand the relationship between trade and environment in this example. You may find it helpful to use the matrix you have developed.

4.2 The global trade in grain

Despite efforts to liberalize the trade in grain, the production of grain is subsidized to a high level within the USA and Europe (one of the results of this is shown in Figure 5.10). These subsidies are seen to distort global trade because they allow producers to offer grain at a price well below the cost of production.

Figure 5.10 EC grain store with its huge grain surplus.
Source:Third World Resurgence, 100/101, 1999, p.29.

Moreover, these subsidies also have a distorting effect on the relationship between trade and the environment. Most of the subsidies go to large-scale farms. This makes it most likely that the subsidies will be used to support intensive agricultural methods. Intensive agriculture needs large inputs of fertilizers, pesticides, and herbicides (see **Morris, 2003**). Often they use so much that the ecosystems cannot absorb the pollution levels that are produced. Fertilizers and pesticides leach into groundwater or surface water and affect water quality. Also, intensive farming methods depend on irrigation, putting additional pressure on local water supplies. Agriculture consumes 70 per cent of the freshwater withdrawn annually by humans. Intensive agriculture is draining more water than is being replenished by rainfall, causing the water table to fall in many regions. Moreover, the heavy use of irrigation can cause salinization of the soil. Indeed salinization is estimated to reduce farm income by 11 billion US dollars each year (French, 2000).

In addition, the subsidies can have a detrimental effect on the social structures in the countries that import the cheap grain (see Figure 5.11). Local farmers cannot compete with the price of cheap imports and many go out of business. One report from the Philippines shows that on the island of Mindanao half a million corn farmers were forced to compete with imports from the USA even though the US corn was subsidized by an amount one hundred times their income (Watkins, 1999).

If the small farms disappear it can also mean a loss of local environmental knowledge. Small-scale farmers often use production methods that are sensitive to the local environment (Figure 5.12) – if for no other reason than because they do not have the money to buy expensive machinery, or the pesticides or fertilizers that can cause so much environmental damage.

In addition, the displacement of small farmers brings other environmental problems as people move onto more marginal lands or into already crowded cities. The 1992 World Bank Development Report provides a good summary of the environmental problems resulting from this kind of displacement:

> Because they lack resources and technology, land-hungry farmers resort to cultivating erosion-prone hillsides and moving into tropical forest areas where crop yields on cleared fields usually drop sharply after just a few years. Poor families often have to meet urgent short-term needs, prompting them to 'mine' natural capital through, for example, excessive cutting of trees for firewood and failure to replace soil nutrients.

> (World Bank, 1992, p.7)

Unlike roses that are a luxury product, the global trade in grain deals in something that is needed for human sustenance. For this reason, the way grain is traded is likely to be more important for the human environment. It also means that issues of power within global trade relations and in the relations between global trade and the environment take on a greater significance. The example of grain therefore provides a link into the discussions about the role of power in the next section.

Figure 5.11 Maize imports being unloaded in an African port.
Figure 5.12 Small-scale farming – traditional agriculture in the Third World.
Source: Third World Resurgence, 100/101, 1999, p.21 and p.28.

Summary

- The relationship between global trade and the environment is understood differently by liberal economists, environmental reformists and radical environmentalists.
- All aspects of the relationship between global trade and the environment are contested.
- How you understand the relationship between global trade and the environment will shape your response to these contests.
- No single perspective provides a complete understanding of the relationship between global trade and the environment: criticisms can be made of each of the perspectives.
- One criticism of all of these perspectives is that they do not directly confront the issues of unequal development.

5   Who controls global trade?

The discussion in the previous section suggested that power may be an issue in the contests between trade and environment. However, the importance of power is itself contested. If you find yourself agreeing with the liberal economic perspective you would play down the importance of power. For example, you would see global trade in roses as the inevitable result of Colombia's comparative advantage with the market controlling what is traded, how much and between which countries. On the other hand, if you took an environmentalist position (reformist or radical) you would think of unequal power relations as an important issue. Furthermore, if you were to pursue the questions raised by a developing country perspective, then power relationships would be integral to your understanding of the trade and environment issue, particularly those between developed and developing countries.

5.1 What form does power take?

If, for the sake of argument, we accept that power may be significant, the next question is: how do we understand the form that it takes? To understand the role of power in more detail we can build on the discussion in **Hinchliffe and Belshaw (2003)** of the two ways in which power works: power resources and power discourses (see also Chapter One for a review).

According to Hinchliffe and Belshaw, power resources relate to the ways in which sources of power can be built up or linked together – the assemblage of power. These are all quite direct forms of power that actors can use to impose their will. It may be physical force or economic superiority that can be used to compel or coerce others to do what they otherwise would not have done. Alternatively, it may come from the backing of powerful institutions, for example the WTO, which adds authority to those who promote liberal economic values. Power discourses relate to the power of ideas and habits. These are much more indirect forms of power that gain the compliance of others by controlling how an issue is understood, for example by controlling whether or not the relationship between trade and environment is discussed or, if it is, how the trade and environment issue is defined.

Building from an understanding of assemblage and discourse, our analysis of the relationship between trade and environment has thrown up another important dimension: the relationships between political, economic and social power. For example, in terms of assemblage some actors are economically powerful but politically weak while others may be politically powerful but economically weak. Alternatively, it may be that they have access to the social power of persuasion, through discourses, but little or no access to direct political or economic power resources. NGOs may be a good example of this. Let's use our examples of roses and grain to look in more detail at how power shapes the contests over trade and environment.

Activity 5.9

What do you think are the different types of power relations operating in the case of our example of the global trade in roses? Who has assembled the most power resources? Is it economic, political or social, and how do they exercise this power – directly or indirectly?

Comment

You might have thought of some of the following points:

- The large rose growers (the TNCs) have built up direct economic power and they use it. In particular, they exercise it over their employees by controlling their working (and social) conditions. Also, because of their substantial contribution to the Colombian national economy, they can demand favourable tax and trade conditions.

- Where the interests of the rose growing companies overlap with the economic interests of the Colombian government there is a powerful combination of economic and political power that supports the prevailing discourse. In other words, both want the current trade to carry on unchanged. This overlap of political and economic interests takes precedence over alternative views that emphasize the social and environmental problems.

- If the views and interests of the Colombian government differed from the TNCs, the Colombian government has substantial political resources that it could exercise directly. However, the government is constrained by its weak economic position and dependence on the economic benefits brought by the rose growing companies.

- The consumers and importing countries also have access to direct economic power – they can choose whether or not to buy the flowers. However, these choices are usually seen as part of the consumers' power of discourse. Suppliers respond to consumer habits and practices: what flowers are wanted and when. Up to the date that this chapter was written, consumers had only used their power to try to alter the relationship between trade and the environment in a few isolated cases (as mentioned by the reformist environmentalist in the dialogue in Section 4).

Now think about the example of the global trade in grain.

- The large agribusinesses (TNCs) have built up direct economic power because they can export their produce at a price that undercuts their local competitors.

- The developing country governments cannot easily assemble power resources – whether economic, political or social – because they are constrained by their reliance on cheap food imports to meet the needs of their populations.

- The developed country governments, in particular the USA and European Union member countries, use their large resources of political and economic power to dominate the trade. Others, who oppose the use of such large agricultural subsidies, are not powerful enough to present a successful challenge.

These notes are just indications, not a definitive list. Maybe you have been able to think of other examples of the use of power? What this exercise shows is that we can use the understanding of power that you have been developing through this series to build up a clearer picture of unequal abilities to influence how resources are allocated and what production processes are used. This can help us understand what happens when the different priorities of trade and the environment clash. However, to round off this chapter we need to return to our original example – the demonstration at Seattle. How can our understanding of power help us to make sense of what was happening there?

5.2 Power and the policy-making process

We already know that the way global trade is currently managed can be traced to a liberal economic discourse. The emphasis is for greater trade liberalization, the benefits of economic growth, and the belief that 'free trade' is the best way to achieve this growth. The agreements and rules of the WTO support this. What we need to find out is how the power relationships maintain this situation.

Around the negotiating tables at Seattle were representatives from most of the world's governments: all of the developed countries and most of the developing countries. There were also meetings scheduled outside the formal conference to discuss matters of concern to NGOs. Beyond these meetings there was a series of large-scale street protests. Not everyone at Seattle had the same capability to influence the discussions. Box 5.3 sets out some key information about the power resources of developed and developing countries, TNCs and NGOs. By examining these we can construct a picture of the power relations that support the current management of global trade and offer some thoughts about the protests over trade and environment at Seattle.

These facts and figures suggest that developed countries have the largest economic and political resources in terms of influencing policy decisions. Because developed countries are the main participants in international trade they will always be directly involved in negotiating trade agreements and formulating regulations. Moreover, the negotiations are conducted in their native languages and according to procedures with which they are familiar.

In contrast, developing countries have very weak economic and political resources. They find it difficult to maintain a constant diplomatic presence; their economic contribution to global trade is significantly less than developed countries; and their bargaining power is weakened because their own national economies are often heavily reliant on global trade with their previous colonizers – the developed countries. Even when developing countries are included in the policy-making discussions, developed countries will have the capabilities to dominate these discussions.

However, because developing countries are members of the WTO they may be able to exercise indirect political power either by exploiting any possible divisions between developed countries over specific issues or by bargaining their compliance on one issue for concessions on another.

Transnational companies are not members of the WTO so their political resources to exercise direct power are low. However, they possess enormous economic and technical resources and many governments, including those of the USA and the EU, often refer to them for their expertise in technical subjects. In addition, developed country governments in particular value their contribution to the domestic economy. This practically guarantees that their concerns will be listened to at a national level and carried forward to negotiations within the WTO.

Box 5.3 Power resources

Developed countries

- Developed countries account for most of the global trade (82 per cent) (United Nations Development Programme 1999).

- The WTO and other major international economic organizations (IEOs), such as the World Bank and International Monetary Fund, are based within developed countries.

- Most of the documents produced by the WTO and major IEOs are published in European languages only.

- The formalities of the WTO are based on the culture of western Europe and the USA. Negotiations are conducted in a way that closely resembles the legal and diplomatic procedures of the West.

- Developed countries can afford to support a diplomatic mission at the WTO, even though this is estimated to be US$900,000 a year according to the UK government.

Developing countries

- Developing countries account for less than a quarter of world trade. Moreover, even this is not shared out evenly: the combined trade of the African countries is only 1 per cent of the total world trade.

- The historical legacy of colonialism has left most developing countries with a weakened economic capacity, poorly developed government structures and underdeveloped social structures (Khor, 2000).

- Many developing countries are still reliant on the markets in the north.

- Thirty countries in Africa and eighteen in Latin America still rely on the export of a few primary commodities for their export earnings (Watkins, 1995).

- The cost of maintaining an extensive diplomatic presence is prohibitive for most developing countries. They may have only one representative or they may share representation with other countries.

- There are not the same strong bonds of shared culture among developing countries that are evident among western developed countries. In negotiations, developing countries rarely act as a homogeneous block.

Transnational companies

Table 5.2 shows that in 1997 the top eleven companies had sales worth more than the Gross Domestic Product (GDP) of many countries. (GDP is the value of all the goods and services produced domestically in one year.)

Table 5.2   Top corporation sales and GDP of countries

Country or corporation	GDP or total sales (US$ billions)
General Motors	164
Thailand	154
Norway	153
Ford Motor	147
Mitsui & Co.	145
Saudi Arabia	140
Mitsubishi	140
Poland	136
Itochu	136
South Africa	129
Royal Dutch/Shell Group	128
Marubeni	124
Greece	123
Sumitomo	119
Exxon	117
Toyota Motor	109
Wal Mart Stores	105
Malaysia	98
Israel	98
Colombia	96
Venezuela	87
Philippines	82

Source: Forbes Magazine, 1998, in UNDP Human Development Report, 1999, p.32.

Transnational companies are becoming bigger as they merge with other companies. Between 1990 and 1997, the number of mergers and acquisitions more than doubled from 11,300 to 24,600. We now have huge, diverse companies that dominate certain sectors. For example: just two companies – Arthur Daniel Midland and Cargill – control the global trade in grain (Juniper, 2001, p.9). There is a similar picture for other commodities. Three companies control 83 per cent of the global trade in cocoa and three companies control 80 per cent of the global trade in bananas (Christian Aid, 1999, p.10).

Non-governmental organizations (NGOs)

- Most NGOs run on a limited budget with relatively few personnel.
- NGOs have little or no access to the formal meetings of the WTO.
- NGOs have only restricted access to the official documents produced by the WTO.
- Very few NGOs can claim to have large-scale popular support. Structures for democratic accountability in NGOs are very weak, if they exist at all.

Moreover, the long-term agenda of TNCs and most governments is similar – they share a commitment to the liberal economic goals of liberalized trade and economic growth achieved through free trade. This means they share the same broad ideas and can expect the current structures of global trade to protect these.

Environmental NGOs have few political and economic resources. They have only recently been acknowledged within the negotiation of trade agreements and regulations. NGO access to documents and processes of negotiation is better than it was but is still extremely limited. Their marginal role has made it easy to dismiss them. However, NGOs are beginning to build up their influence in terms of presenting alternative discourses. The demonstrations outside the conference centre, and the meetings held to coincide with the ministerial meetings, show that there is an increasingly vocal body of opinion. Moreover, the expressions of concern coming from NGOs appear to be having some kind of an impact. For example, with street protests still disrupting the ministerial meeting, President Clinton speaking inside the negotiations urged delegates to listen to 'legitimate concerns of legitimate protesters' (*The Guardian*, 1999, p.1).

Nevertheless, NGOs are still hampered by their diversity. The myriad of groups hold different priorities and adopt different approaches. Perhaps it will be easy for the WTO to provide concessions for one or two issues that will dissipate the NGOs forceful opposition without significantly altering the basic liberal economic agenda.

5.3 How do the power relations maintain the liberal economic agenda?

Now that we have looked at the power capabilities of the main players, let's go back to our earlier question: how do the power relations maintain the liberal economic agenda? Clearly the power relationships among those at Seattle were not equal. Developed country governments are more powerful in terms of their assemblage of resources. They also exercise discourse power by controlling the agenda using persuasion, manipulation and negotiation. Moreover, the authority of international institutions, in particular the WTO, supports the developed country position. Developing countries are significantly weaker and have far fewer resources to support direct power. Moreover, they cannot rely on international institutions to support their position. This lack of power resources is reflected in their lack of influence in shaping the agenda within the WTO. On a broad level, they share the developed countries' commitment to a liberal trade agenda. However, developing countries have their own particular concerns, especially the trade barriers in developed countries and the use of government subsidies in developed countries. The non-governmental organizations – TNCs and environmental NGOs – seek to influence their own governments and the more powerful developed country governments. However, TNCs have greater economic power to achieve this. Moreover, TNCs also generally support a liberal economic agenda and so, combined with their direct economic power, they have some influence in shaping the trade agenda. As the players with the least power resources, the NGOs will find it difficult to overturn the liberal agenda. Their demonstrations for radical change can be physically quashed or just ignored and their more passive attempts to press for change through meetings and persuasion can be managed so that they do not upset the status quo. However, environmental NGOs do have some power of discourse and the demonstrations are an indication that they are becoming more successful in organizing a body of common ideas to challenge the agenda going on inside the WTO. As a way of reflecting on the events in Seattle, you might like to think about the following question.

○   Using your knowledge and understanding of the trade and environment issue, how likely do you think it is that NGOs will be part of a movement that will change how trade is conducted?

●   There is no right or wrong answer, but think carefully. The flip side of the question is to ask yourself this: do you think the global trading system will always be the same?

Summary

- Not everyone agrees on the significance of power for understanding the relationship between trade and the environment.
- It is useful to think of power in terms of the material capabilities to exercise direct power, the authority of international institutions, and the indirect power of ideas to control the agenda.
- The contests over trade and environment add a further dimension: the distinction between economic, political and social power.
- Those who are able to exercise a strong influence on how trade is practised are powerful in more than one area.

6   Conclusion

In a book about contests over environmental issues, the main aim of this chapter has been to give you a chance to look at an issue where these contests have a global dimension – the relationship between global trade and the environment. One graphic example of these contests is the demonstration outside the WTO Ministerial Meeting at Seattle in 1999, that called for an overhaul of the current trading system to take more account of the human and non-human environment.

This chapter has used an investigation of why the demonstration took place and what was at stake, to deepen your understanding of the book's three key questions:

- How and why do contests over environmental issues arise?
- How are environmental understandings produced?
- Can environmental inequalities be reduced?

The chapter has demonstrated that the contests are also about different understandings of the trade and environment issue. A number of important distinctions have been made between the liberal economic, environmental reformist, radical environmentalist and developing country perspectives.

At an elemental level, there is no agreement that trade does have a direct effect on the environment. Even when the effects – whether direct or indirect – are considered, there is disagreement about whether, overall, the current system of global trade is harmful or beneficial to the human and non-human environment.

However, by applying these views to real examples the chapter has shown that there is no conclusive evidence to support one view over the others. You should now be able to identify these views and critically assess them for the accuracy of their assumptions, the relevance of their approach and the feasibility of their prescriptions. Moreover, you will be able to use these perspectives to critically analyse examples where the priorities of global trade and the environment come into conflict.

By reflecting on the different perspectives we were led to consider the relevance of power. Although the assumptions behind a liberal economic view downplay its significance it remains an important consideration for environmentalists. This chapter has been able to use the understanding of power that you have built up during this series. To make it even more pertinent we extended this definition to identify economic, political and social power. This proved useful for explaining exactly what was happening at Seattle. Furthermore, by using the 'snapshots' of power resources you should be able to discuss how the assemblage of power resources and power discourses at Seattle explains how the contests are played out. It has also helped us to understand why one view – the liberal economic – prevails over the others.

Through your understanding of what happened at Seattle you should now be able to apply these ideas to other demonstrations and conflicts between global trade and the environment.

References

Castellanos, J. (1997) *Rose Trade and the Environment*, TED Case Studies, www.american.edu/projects/mandala/TED.

Castree, N.C. (2003) 'Uneven development, globalization and environmental change' in Morris, R.M. et al. (eds).

Christian Aid (1999) *Fair Shares? Transnational Companies, the WTO and the World's Poorest Communities,* London, Christian Aid.

Ferrer, Y. (1997) 'Women and agriculture: Colombia's roses prove a thorny issue', *Bogota*, International Press Service, cited in http://www.american.edu/projects/madala/TED/rose.htm (accessed 24 August 1997).

French, H. (2000) *Vanishing Borders: Protecting the Planet in the Age of Globalization*, London, Earthscan.

GATT (General Agreement on Tariffs and trade) (1992) Report on Trade and the Environment, Geneva, GATT Secretariat.

Held, D., McGrew, A., Goldblatt, D. and Perraton, J. (1999) *Global Transformations: Politics, Economics and Culture,* Cambridge, Polity Press.

Hinchliffe, S.J. and Belshaw, C.D. (2003) 'Who cares? Values, power and action in environmental contests' in Hinchliffe, S.J. et al. (eds).

Hinchliffe, S.J., Blowers, A.T. and Freeland, J.R. (eds) (2003) *Understanding Environmental Issues,* Chichester, John Wiley & Sons/The Open University (Book 1 of this series).

Juniper, T. (2001) 'Rich pickings', *The Guardian G2*, pp.8–9.

Khor, M. (2000) *Globalization and the South: Some Critical Issues*, Discussion papers, United Nations Conference on Trade and Development, Geneva.

Maharaj, N. and Dorren, G. (1995) *The Game of the Rose: The Third World in the Global Flower Trade*, Utrecht, International Books.

Morris, R.M. (2003) 'Changing Land' in Morris, R.M. et al. (eds).

Morris, R.M., Freeland, J.R., Hinchliffe, S.J. and Smith, S.G. (eds) *Changing Environments,* Chichester, John Wiley & Sons/The Open University (Book 2 of this series).

Public Citizen (2001) http://www.citizen.org/trade/wto/shrink_sink/ (accessed 16 January 2002).

United Nations Development Programme (1999) *Human Development Report*, New York, United Nations Development Programme.

United Nations Research Institute for Social Development (1995) *States of Disarray: The Social Effects of Globalization*, Geneva, United Nations Research Institute for Social Development.

Watkins, K. (1995) *The Oxfam Poverty Report*, UK and Ireland, Oxfam.

Watkins, K. (1999) 'Free trade and farm fallacies', *Third World Resurgence,* no.100/101, pp.33–7.

Watkins, K. (2001) 'Deadly Blooms', *The Guardian*, 29 August 2001.

Williams, M. (2001) 'In search of global standards: the political economy of trade and the environment' in Stevis, D.S. and Assetto, V.A. (eds) *The International Political Economy of the Environment*, London, Lynne Rienner Publishers.

World Bank (1992) *World Bank Development Report: Development and the Environment*, Washington, World Bank.

World Trade Organization (WTO) (1999a) *Trade and Environment*, Special Studies, Geneva, WTO.

World Trade Organization (WTO) (1999b) Press release on Special Study on *Trade and Environment*, www.WTO.org/english/tratop_e/envir_e/stud99.htm (accessed 18 April 2001).

WWF (1999) *Report No.2* from the WTO Ministerial Meeting in Seattle, www.panda.org/resources/publications/sustainability/wto-papers/seattle-report2 (accessed 22 January 2002).

Chapter Six

Environmental justice and the environmental justice movement

Wendy Maples

Contents

1   Introduction

The 'alternative national anthem' of the United States of America shows the extent to which America's history and Americans' sense of national identity are tied to the land. Notable omissions from the song are the cityscapes of the cosmopolitan coasts and industrial centres, the backwaters of the South and the dry plains of middle America; instead, the scene is solely one of plenty, abundance – and, importantly, 'brotherhood'.

This idealized image of the USA, illustrated in Figure 6.1, is one to which many Americans feel a close affinity. However, for some Americans, while the ideal persists, the reality has become so derelict the song might seem a mockery of their everyday lives. For some, indeed, the very environment that supposedly sustains and gives Americans a sense of national identity has been so badly polluted as to be a source not of sustenance but of illness, disease and death.

The 'brotherhood' of the American natural idyll has also come under question. Some research suggests that the pollution that corrupts the water, soil and air – and therefore the everyday lives of many Americans – is not shared equally among the brethren. Instead, it is argued, the effects of environmental ills, such as toxic waste dumps, particulate-producing factories and routine mass pesticide sprays, are disproportionately suffered by the poor, disadvantaged, politically disenfranchised, women and ethnic and racial minorities (see for example: Bullard, 2000; Pulido, 1996; Rothman, 1998; Szasz, 1994).

Throughout US history, socio-economic affluence has been linked to a rise in the use of new technologies and greater consumption of disposable goods by society's wealthier members – and to a corresponding increase in pollution, land degradation and depreciation of local environments, predominantly in areas inhabited by the poor. So, for example, there are freeways that tend to run through or alongside more poor than wealthy neighbourhoods, and carry cars and goods disproportionately benefiting middle class workers and consumers. This inequity in environmental amenity – availability of environmental 'goods' and imposition of environmental 'bads' – has become a spur for action in the USA. The specific contest about environmental inequality which traces divisions along lines of social, economic and political disparity is the raison d'être for what has become known as the **environmental justice movement (EJM).** The EJM emerged from the political action of people concerned to address environmental issues also as issues of social justice.

environmental justice movement (EJM)

This chapter looks at the development of the EJM from its roots in the experiences of underprivileged US groups, and the values associated with that development. Although this means that the chapter mainly focuses on the environmental political history of the USA, the concern with environmental inequality now has resonance internationally. Indeed, wherever you are in the world, you are likely to know of a local environmental dispute that will have involved some sort of political intervention. At the heart of many such

O Beautiful,
For spacious skies
For amber waves of grain
For purple mountains, majesty
Above the fruited plains
America, America
God shed his grace on thee
And crown thy good,
With Brotherhood
From sea to shining sea!

Figure 6.1 'America the beautiful': the American natural idyll.

interventions is the question of what is fair, tolerable or equitable for the local people – the ideas or principles of 'environmental justice'. In order to emphasize the broader dimensions of actions now taken in the name of environmental justice, while two of the three case studies of political conflict to which you will be introduced in the chapter relate to events in the USA, the other involves incidents in Nigeria.

Overall, the chapter considers how the distribution of environmental 'goods' and 'bads' might be related to other political, economic and social inequalities, and to explore how the EJM came to influence mainstream environmental values and action. Central here will be the initially conflicting conceptualizations of the environment held by the more powerful mainstream environmentalists and those held by so-called 'subaltern' groups. This conceptual division over the environment will lead us to think more about the operation of power. In particular, we will encounter power being understood and utilized not only in terms of resource but also discourse (see Chapter One and also **Hinchliffe and Belshaw, 2003**). As you should recall, this distinction suggests power rests not only with institutions and formal activities, but is also internalized and integral to people's everyday lives, to how we think and behave. The chapter aims, therefore, to deepen your understanding of the political nature of environmental disputes, to encourage you to engage in some critical reasoning on different dimensions of environmental justice, and to help you further reflect on your understanding of the connections between environments, values, power and action.

To help achieve these aims, we will draw on the book's key questions. First we might ask: why do contests over the environment arise? I will break with convention here and provide a partial response right now. According to the EJM, contests arise because of environmental inequalities. As I am sure you realize by now, this is not the whole answer. We might ask further then: why are there environmental inequalities? This in turn leads us to another question: how might environmental inequalities be reduced? The EJM does offer answers to these questions, but it will be up to you to decide whether or not they are convincing. Finally the chapter addresses the question of how environmental understandings are produced – in this case how the EJM has influenced the way that mainstream environmental organizations conceptualize the environment.

2   The beginning of the environmental justice movement

A good place to begin our exploration of the notion of an environmental justice movement is with the story of Love Canal, a residential district of Niagara Falls in upper New York state, where for many US writers and activists the EJM began.

At one point in its history, Love Canal was a partially completed, then abandoned hydroelectric canal. It subsequently became a dump for, reportedly, 'thousands of drums' of toxic, chemical waste (including chloroform, benzene, toluene and trichloroethylene) deposited by Hooker Chemicals, which had a plant adjacent to Love Canal (see Blowers 1996). In 1952 Hooker Chemicals covered over the dump site. In 1953 the site was sold to the Niagara Falls Board of Education. Despite warnings from Hooker about the chemical deposits, a school was built and homes were developed on the surrounding land. (See Figures 6.2 and 6.3.)

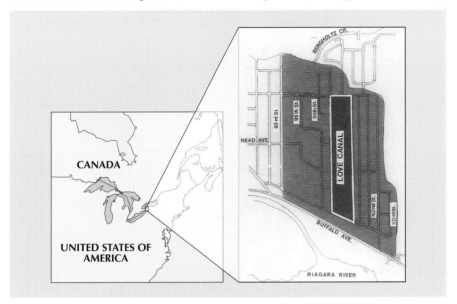

Figure 6.2 The location of Niagara Falls and street map of the Love Canal built up area; the school was built on 99th Street.
Source: Rothman, 1998, p150.

As historian Hal Rothman describes it:

> An ordinary neighbourhood took shape. It was largely made up of chemical and industrial plant workers and their families, typical of Niagara Falls in their economic situation, educational level, and almost every other form of demography ...

Figure 6.3 Aerial photographs of the Love Canal area: (left) an early photograph showing Hooker Chemicals plant complex sited close to the water, and (right) in 1978, showing the school in the centre and surrounding homes.

Until the early 1970s there seemed to be nothing terribly unusual about the Love Canal neighbourhood. In 1958, three children were burned by exposure to chemical residues on the surface of the canal, but no one detected any pattern of danger. Children played in the debris atop the filled-in canal. Chunks of phosphorus, which children called 'fire rocks', were favoured; thrown against cement, they would explode, sending off a trail of white sparks ... [N]o one seemed terribly concerned. Love Canal was only one of many lower-middle-class communities, populated with the families of first-generation homeowners indebted to the paint, chemical, pesticide, and related industries that gave them the opportunity to achieve the American Dream.

(Rothman, 1998, pp.149–51)

Eventually, patterns of illness developed that became too obvious to ignore, yet a series of denials from local government meant there was no official public acknowledgement of the problems. The EPA (the US Federal Government's Environmental Protection Agency) was alerted by city officials to the toxics risk in 1977. Concerned that the canal dump site had been disturbed during the period of construction and that there might be seepage into the soil and groundwater, the EPA advocated evacuation of local residents. Nothing, however, was done to achieve this for nearly a year. Only the tenacity of local homeowner Lois Gibbs and journalist Michael Brown ensured the concerns of the community – and information about local toxicity – were brought to the attention of a wider public through demonstrations, letters to the papers and various publicity campaigns (see Figure 6.4). And only then did state Health Commissioner Robert Whalen address the issue publicly, stating what residents had suspected: Love Canal was a 'great and imminent peril to the health of the general public' (Rothman, 1998, p.151). In a matter of months the area was evacuated (Figure 6.5) and then fenced off, although the chemical seepage in soil and groundwater continued unabated.

While many evacuated residents reported 'feeling better', relocation away from the area did not fully halt the illnesses. The chemical compounds to which they had been exposed led to skin conditions, high rates of hearing impairment and

Figure 6.4 Love Canal residents' protest.

Figure 6.5 One of the abandoned houses in Love Canal, following the evacuation of the residents.

Source: Rothman, 1998, pp.152 and 154.

cancers, as well as chromosomal damage which had the potential to cause 'cancer, miscarriage and birth defects' (Szasz, 1994, p.86).

The Love Canal Homeowners Association was set up to protest collectively for some settlement to the residents' grievances. The Love Canal story was widely reported, with national news coverage resuming whenever new angles on the tragedy were brought to light. Moreover, it was made plain in many reports that Love Canal was only the tip of a toxics iceberg.

2.1 Love Canal: a success story?

According to Lois Gibbs (who later became organizer of the Citizens' Clearing-house for Hazardous Waste), there were many factors that meant her community was let down. There was the environmental disregard of Hooker Chemicals, the apparent complacency of local government, the lack of intervention by an evidently toothless EPA and the unwillingness of the population to believe that industry and government would allow its citizens to be poisoned. Trying to get 'facts' from state officials was all but impossible and getting fair compensation proved just as difficult. The residents' struggle involved learning to understand the technical vocabulary used by various officials and experts, pulling together the evidence (often obscured from public view) generated by scientific reports, and gaining help from sympathetic legal advisers, the local press and, later, national news coverage. While in some regards Love Canal is a success story – official recognition of the problem was achieved, people were rehoused and some compensations paid – the gains made by the residents were only possible due to the enormous publicity and to the political tenacity of the residents themselves.

Activity 6.1

Lois Gibbs claims that: 'The moral of the story is that the only way we can fight these situations is by fighting politically' (quoted in Edwards, 1995, p.48).

What do you think Gibbs means by 'fighting politically'? Make a few notes on whether you agree or disagree with her claim.

Comment

Writing letters and articles for the local press, presenting evidence to local politicians and pressing for action from the EPA are all examples of fighting politically. They are examples of power in action – where the concerned parties use a range of resources to meet their objectives. In this case, the residents made use of public, scientific documents, personal testimony, doctors' reports, the assistance of sympathetic lawyers and the media to push for attention, compensation and change. Gibbs goes on to say: 'We cannot fight scientifically alone. We cannot fight legally alone' (quoted in Edwards, 1995, p.48). In the case of Love Canal, there was plenty of scientific evidence to support the fear that seepage had occurred, but it was not until a national spotlight was cast upon the case that many families were rehoused. While scientific and legal battles are not dismissed by Gibbs, she feels that the scientific evidence and the legal system must be used *politically* to be effective.

But was the polluting of Love Canal anything more than a localized environmental disaster? On a grander scale, the emerging evidence was that there were potentially tens of thousands of 'Love Canals' across the USA, and this led to the creation of the 1979 'Superfund' Act (legally known as the 'Comprehensive Environmental Response, Compensation, and Liability Act'), a US$1.6 billion fund set up with federal taxes on petroleum companies and other potential polluters to support toxic clean-ups.

But there is another sense in which Love Canal is more than just an isolated case. This chapter began with the promise of a discussion of environmental justice, and I claimed that Love Canal marked the beginning of the EJM. Now is a good opportunity to describe in more detail what is meant by environmental justice and the EJM.

Summary

- The story of Love Canal is one of an ordinary neighbourhood, where the disregard for environmental pollution and its effects on local people led to extensive protests and campaigns and which led finally to the evacuation of residents.
- Love Canal is one of the first success stories in the recognition of the need for environmental justice.
- The case study provides an example of power – in this case political power – in action, and of power exerted by people.

3   Environmental injustice and the environmental justice movement (EJM)

It is stating the obvious to note that human habitation of environments is not a uniform experience: in some places the streets are unswept, or the air is heavy with traffic fumes, or pesticides hang in the air and cling to clothes and skin, or industrial waste lurks beneath the soil; while in altogether different areas, none of these problems is evident. This *uneven* distribution of both environmental 'goods' and 'bads' is a key aspect of environmental justice concerns and is, therefore, worth emphasizing. Importantly, say EJM advocates, the uneven distribution of environmental 'goods' and 'bads' correlates with demographic variables of economic wealth, racial and ethnic difference and political clout.

According to the EJM, there are important reasons why the *relatively* poor, lower middle-class residents suffered from the toxic cocktail of Love Canal. Like the nuclear communities discussed in Chapter Three, the Love Canal community was tied to the very industry that polluted it: they were employees of the nearby chemical, pesticide and paint industries. In a sense, Love Canal was a 'factory town' where housing was affordable for low-to-middle income families.

This is significant because, according to Andrew Szasz, low-to-middle income families living in factory towns tend to be politically conservative and to believe that employers and the government look out for the health and safety of their people. He refers to a poignant question (one that lies at the heart of any enquiry into environmental injustice) asked by Lew Dunn, who lived near a hazardous waste disposal site in Casmalia, California, and became active in anti-toxics campaigning when it was apparent that the site was leaking toxic wastes into the local groundwater and that the local authorities were not prepared to do anything about it. That question was 'Why don't I see a toxic waste dump in Beverly Hills or next to the governor's mansion?' (quoted in Szasz, 1995, p.98). Szasz responds by arguing that:

> The burden of living with industrial waste has always been unevenly distributed, falling more heavily on poor than on well to do, more heavily on black and brown than on white. Although deplorable, at least this used to be, in a sense, an 'unconscious' practice. It was the inevitable, though not necessarily intended consequence of normal business logic of choosing the least expensive option in a regulation-free business environment. Now, however, some [corporations have] advocated a *conscious* policy of identifying those communities least able to resist.
>
> (Szasz, 1994, pp.104–5)

The concern is that siting criteria may thereby prioritize least resistance over geological safety. A 1984 study conducted for the California Waste Management Board suggested that a prime concern of site locators *should* be the potential receptivity of inhabitants. According to the findings of investigating consultants, people who were least resistant to LULUs (locally unwanted land uses) could be typically identified as those who:

> live in a rural community or small community of fewer than 25,000 people in the South or Midwest; are 'old-timer' residents, are low income; are ranchers, farmers, or work in areas that are business related, technology related, or nature exploitative; have a high school education, or less; are above middle age; are politically Republican, conservative, and have a free market orientation; are not concerned about the environment; are not politically active or involved in voluntary associations; and are Catholic.
>
> (Szasz, 1994, p.105)

The discovery of increasing numbers of 'Love Canals', with their implicit and, in some cases, clearly explicit discriminatory policies, coupled with a culture that encourages seeing conflicts in terms of citizens' or civil rights, led to a redefinition of 'the environment' and 'environmentalism' in the USA in the 1970s and 1980s.

Such gains and shifts associated with what we now term 'environmental justice' is not meant to imply that the USA 'invented' a concern with the poor and their environmental conditions. In another century and on another continent, to take just one example, Charles Dickens wrote about precisely this connection in his evocation of 'Coketown' in the novel *Hard Times*:

> It was a town of machinery and tall chimneys, out of which interminable serpents of smoke trailed themselves for ever and ever, and never got uncoiled. It had a black canal in it, and a river that ran purple with ill-smelling dye, and vast piles of building where the piston of the steam-engine worked montonously up and down, like the head of an elephant in a state of melancholy madness.

> (Dickens, 1854, p.10)

Indeed Dickens, like the EJM much more recently, recognizes not only the problems of Coketown itself, but also the relative injustice of the situation: a Coketown existence is compared to the world of those rich enough to buy the goods produced there and who 'could hardly scarcely bear to hear the place mentioned' (Dickens, 1854, p.10). If not dealing with entirely new issues then, what is nonetheless of lasting significance and relevance about the EJM is the way that it brought together for the first time environmental questions (arising in this case from particular histories of Americans' relationships with their natural and human surroundings), and issues of civil and human rights which had been pushed onto the agenda in the USA since the mid-twentieth century.

From the civil rights movement in particular came a focus on how to redress the consequences of established social inequalities. Protests became part of a process designed to highlight the extent of inequalities brought about by racial discrimination. The result of such protests, it was hoped, would be legislation that condemned and eventually eradicated such discrimation throughout the USA. For environmental justice advocates, similarly, legislative change came to be seen as essential. But what would the EJM hope new legislation might do? If the goal is to acheive some kind of equitable justice, what would that mean? In some cases, the EJM has advocated legislation that ensures distributive justice; in others, procedural justice has been the goal. We will now turn to look at these two approaches in more detail.

3.1 What constitutes environmental injustice?

Put simply, environmental injustice is the inequitable or uneven distribution of environmental 'bads' or 'goods'. Environmental injustice understood in these terms has one solution: 'distributive justice', the redistribution of environmental 'bads' (for instance, hazardous wastes to the governor's mansion) and 'goods' (for instance, improved access to local beauty spots through provision of buses that run from poorer areas to the beach). Although distributive justice is more complex than this (the governor's mansion may well turn out to be geologically unfit for hazardous waste disposal), the basic principle of equitable distribution of both environmental 'bads' and 'goods' is one that US legislators are now encouraged to bear in mind.

Activity 6.2

Have a look at Buckminster Fuller's map of the world, shown in Figure 6.6. There are over 6 billion people on the earth and approximately 26.7 billion acres (10.8 billion hectares) of productive space (excluding, for example, the polar ice caps or desert extremes). Simplistically, an *equal* distribution of productive space (which includes productive ocean as well as productive land) would involve giving 4.41 acres (1.8 hectares) to each person currently on the planet.

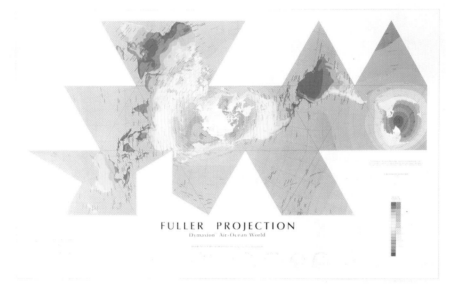

FULLER   PROJECTION
Dymaxion Air-Ocean World

(The map shows mean low temperature for land and water: the bar on the left in the key shows land temperature from below 58° F (the bottom of the bar) to above 77° F; the bar on the right in the key shows water temperature from below 40° C (the bottom of the bar) to above 20° C.)

Figure 6.6 Buckminster Fuller's flat globe provides a good indication of relative land/ water mass.
Source: The Fuller Projection Map design is a trademark of the Buckminster Fuller Institute © 1938,1967 and 1992. All rights reserved.

Now consider: while it might be spatially 'equal', why might it be *inequitable* to distribute the earth in this way?

Comment

One reason why it might be inequitable to divide all the earth's habitable land equally among all of its human inhabitants is that some land is *more* habitable than other land. Some areas have water and some have higher levels of pollutants. Or, you might argue, perhaps not everyone has the same rights to land as others: children arguably have less need of a full share of land. We might also consider the needs of future generations. These are questions of social justice – and they raise questions about who decides who gets what and *according to which principles.*

3.2 Principles of distributive justice

Distributive justice requires consideration of some important principles. First, we need to ask to *whom* to distribute: who is included in the community of justice? We might start by thinking about ourselves, and perhaps our children. I have already mentioned future generations, but the list of those who might be included is yet longer. The second question we need to ask then is: what is the *principle* of distribution? There are myriad answers to such questions. Boxes 6.1 and 6.2 are two lists adapted from research by Andrew Dobson, a researcher of environmental politics and philosophy. Before looking at the first list in Box 6.1, take a moment to consider who (or what) *you* would suggest should be included in the community of environmental justice.

Box 6.1 Who should be included in the community of justice?

- present generation, human beings

- present and future generation, human beings

- present and future generation, all sentient beings (humans and other animals)

- present and future generation, all beings (sentient and non-sentient)

- 'agent-affected' (anyone or thing affected by environmental conditions caused by the actions of another, the 'agent')

(adapted from Dobson, 1998, p.63, Table 5)

Dobson's list shows that identifying those to include is a complex task. But let us move on to consider the principle of distribution. One possibility has already been mentioned – the equal allocation of productive space – a distributive principle that might be *inequitable*. Dobson's next list (in Box 6.2) offers some alternative principles of distribution.

Box 6.2 What is the principle of distribution?

- equality

- utility

- entitlement

- needs

- desert

- to the benefit of the least advantaged

- market value

(adapted from Dobson, 1998, p.63, Table 5)

These may seem quite abstract principles, so spend a minute or two thinking about what each might mean in practice. What would a principle of 'need' for the present generation of human beings mean in everyday practice? It might equate with basic human needs such as water, food, shelter and clothing. What about distribution on the basis of market value? This is easy to imagine if we think about terrestrial property, but what about other elements of our environment such as oceans or rivers or air? Despite the appearance of oxygen bars (Figure 6.7), it is rather more difficult to imagine a market value for air. And what if you wished to account for the value of air to both present and future generations? What about the principle of 'to the benefit of the least advantaged'? This is similar to the social concept of helping those who are unable to help themselves: in Robin Hood style, the priority would be to create equity by mitigating current inequity.

Figure 6.7 Oxygen bars offer customers scented and unscented oxygen in pure quantities for perceived health benefits.

And what would you distribute? If you are distributing land, is it land for agricultural or aesthetic purposes? Are you distributing both environmental benefits and burdens – an equal share of clean water and waste sites, of salinated soil and virgin forest?

Put simply, different priorities may well indicate substantially different outcomes. Should you, for instance, extend the distribution of resources to all present and future generations? An immediate result might be that there is 'less' available in the immediate term. Using the example of water in a stream, you might argue that the water should be cared for in order that it was *in the same condition* for present and future generations. Given the current state of many streams this may well involve a good deal of intervention to prevent the build-up of litter or the flow of polluted water from nearby farms or factories into the stream. So although shares of the stream water are the same for both current and future generations, the burden of ensuring this good might be disproportionately carried by the current generation.

At the present time, the Earth's available resources are unequally and, some would argue, inequitably distributed. However, apart from the extremes, it is difficult to determine who is currently resource poor and who is resource rich, as there are so many variables to take into account. Some people have few material possessions but have quite a lot of land, and if they have land it may have either high or low productive potential. Other people may have vast material possessions, but live in a relatively small space, shared with large numbers of other people. Yet others may have both land and material possessions, but the air they breathe is polluted.

Despite these difficulties, however, there are activists, academics and legal professionals whose concern with the environment and social justice has led them to believe that distributive justice legislation is worth pursuing.

3.3 Principles of procedural justice

As you have seen, equitable distribution is no mean feat, which is why some activists and theorists have argued that it may be better to work towards **procedural justice** rather than distributive justice.

procedural justice

According to Robert Bullard, sociologist, civil rights and environmental justice activist, procedural equity refers to:

> The extent to which governing rules and regulations, evaluation criteria, and enforcement are applied in a nondiscriminatory manner. Unequal protection results from nonscientific and undemocratic decisions, such as exclusionary practices, conflicts of interest, [or] public hearings held in remote locations and at inconvenient times.
>
> (Bullard, 2000, p.116)

Where there are procedural *injustices*, there are flaws in the very mechanisms that are meant to ensure, in the American vernacular, 'justice for all'.

One of the seventeen principles of environmental justice adopted at the First National People of Color Environmental Leadership Summit in Washington DC, in 1991, was the following:

> Environmental justice demands the right to participate as equal partners at every level of decision-making including needs assessment, planning, implementation, enforcement and evaluation.

Many of the other principles adopted also focused on 'rights'. This is not surprising given that many environmental justice proponents in the USA emerged from the civil rights movement. The language of 'rights' is one that has a hearty resonance in the USA. It also lends itself well to a culture that is beholden to the legal profession and statute for the establishment of civil law.

It is, therefore, also unsurprising that success in the environmental justice lobby has sometimes been measured in legislation. To the pleasure of the EJM, in 1994 President Bill Clinton signed Executive Order 12898, requiring 'Federal Actions to Address Environmental Justice in Minority Populations and Low-Income Populations'. The order requires federal agencies:

> to ensure that minority and low-income communities have access to public information on human health and the environment; consider dispropor-tionate health effects when conducting research and collecting data; and conduct activities related to health and the environment in a manner that does not discriminate against minorities.
>
> (*International Wildlife*, 1994, p.26).

Five years later, Congressman John Lewis introduced further legislation under the title 'The Environmental Justice Act'.

3.4 Legislating for environmental justice

US legislators have made some very dramatic moves in an attempt to address pollution problems in the American backyard and to protect the communities most badly affected by environmental injustices.

Executive Order 12898 is an attempt at procedural equity, given its emphasis on access to *information*. This is not an attempt to redistribute 'goods' or 'bads', but to ensure that, for instance, access to information is fair and nondiscriminatory. The Environmental Justice Act is a little more complex. It empowers the EPA to research and provide relief for the most polluted communities in the USA, and the Department of Health and Human Services to investigate connections between environmental pollution and the health of residents in badly affected areas. It is concerned with distributive equity in as much as it requires the EPA to seek redress for previous wrongs. Money is taken from potential polluters in the form of taxes which are used to clean up polluted areas. But the focus of the legislation is the EPA, and as such it could be seen as a corrective to previous EPA policy and as an attempt to achieve procedural as well as distributive equity.

Summary

- The EJM was and is a social and political response to perceived environmental *injustices*. These injustices, argue the EJM, could be addressed through the establishment of legislation that serves to ensure either distributional or, perhaps more usefully, procedural equality.

- Although we have not used this terminology, the preceding section was very much concerned with *values*. The principles of environmental justice were at odds with the values of other groups at the time – legislators, regulators and industry. EJM principles such as those set out by the First National People of Color Environmental Leadership Summit challenged the values of industry and government and their power over the environment of people of colour and other 'peripheral communities'. The principles of environmental justice, then, are linked with the concepts of values and power.

- Justice is understood in different ways – distributively and procedurally; environmental justice emerged in part from the ideas of the civil rights movement: the EJM understandably charts success in terms of legislative gains.

4 'The environment' as discourse

In Section 2, we touched on the way power was understood as a resource. In this section, I want to explore the idea of power as *discourse*. The reason for this is that one of the main challenges faced by the EJM was the predominant assumptions of the mainstream environmental movement. Many EJM campaigners complained that mainstream environmentalists effectively ignored the plight of urban and rural people because their interests were seen as outside the purview of 'the environment'.

The environment has different meanings depending on who you are, where you live, what you do, etc. As noted in the introduction to this chapter, how we understand our environment is partially constructed through 'discourse' (see also Chapter One, and **Hinchliffe and Belshaw, 2003**).

Activity 6.3a

Draw up a blank table divided into two columns. Label the first column, 'The Environment'. Leave the other column blank for the time being. Take just a minute or so to note down in the first column whatever images/places come to mind as you think about 'the environment'.

Comment

What came to your mind? My own list includes:

- deer grazing in a small woodland clearing
- a telecoms tower at the very top of a high mountain
- waves crashing onto a rocky shoreline
- a 'micro-park', with a patch of grass and a rock or two
- the South Downs of England – all green lushness and rolling, chalky hills.

(See Figure 6.8.)

Activity 6.3b

Now label the other column 'My Environment'. In this column, briefly write down what you see in front of or beside you as you are reading this book.

Comment

My list includes:

- an unseemly pile of washing up
- some potted plants and an unmown lawn
- a cat in the process of eating a fly (yuck!)
- a pile of firewood
- a road with noisy traffic
- lots of different species of trees

Now, this list would have looked very different if I were writing at the office, or if I were writing 'on the move', as I often am, taking a train to the Open University at Milton Keynes. Nevertheless, this was my immediate environment at the time I compiled my list.

We often have preconceived ideas about the environment, what it means and what aspects of it are significant. Yet, what we *imagine* is significant and what we experience on a day-to-day basis may well differ. Why do you think this is? One way of thinking about it is this: what we perceive as 'the environment' is not natural or given, it is subject to social and cultural biases – ideas we have learned about what our society or culture 'values' about the environment. This is *not* to say that there is no material world out there. Of course, trees, birds, flies, oceans and dirt exist materially. What they *mean* to us, however – their significance and their everyday value – is not fixed but changes over time.

Importantly, too, at any given time some meanings are more powerful or dominant than others. Why, for instance, when my immediate environment is dirty dishes and, now, cat sick, have I started my imaginings about the environment with deer grazing in a woodland clearing? You might have noticed that this question leads to one of the key questions for this book: how are environmental understandings produced? For some theorists, an answer to this lies in 'discourse'. Let me take a few moments to elaborate on why I find the concept of discourse helpful in understanding the rise of 'environmental justice' concerns.

4.1 Discourse

Michel Foucault (1926–1984), a French philosopher, posited that our social and cultural worlds – the experiences we have and the very ideas we carry round in our heads – develop through what he called 'discourses'. The term discourse describes both what is talked about and *how* it is talked about. For Foucault, 'discourse' described better the way that power is exercised than, for instance, 'top-down' versions of power which tended to see power as essentially to do with authority and physical coercion. My deer, crashing waves and South Downs (Figure 6.8) are images borne out of a discourse that imagines 'the environment' as being primarily about places of 'natural beauty'. If I signed up for a university course about the environment, I might not expect to be reading about washing up, or cat sick. But I would expect to learn something about woodlands, the oceans, and the geology that produces hills and mountains.

Figure 6.8 Images of outstanding natural beauty: (a) fallow deer grazing, (b) telecommunications tower for relaying microwaves, set on a mountain top, (c) waves crashing onto the shoreline at the Giant's Causeway in Ireland; (d) a micro park in Los Angeles; and (e) the lush South Downs of England.

The reason for this is that many contemporary discourses about our environment are based on the idea of this environment as being something 'out there', *a part of* the 'natural world', and, often, something *apart from* humans, from 'us'. In this book and the series we have been at pains to look at the interconnectedness between the natural and human worlds. We have felt that we *needed* to adopt this approach very consciously, in part because other, 'segregationist' discourses remain pervasive at the time of writing. The materiality of our world is not in question; the concern of environmentalists is the extent to which some of us live our everyday lives *as if* that materiality were of little consequence.

What is most curious (some would say pernicious) about all of this is the extent to which it seems so 'natural' (and by this I mean it comes to us with ease). The song that began this chapter presents an image of America that is very captivating. But if we look more closely at the images evoked and begin to think about *what is left out*, we may find the words more discomforting. Looking beyond that romanticized vision of 'America the Beautiful' (shown again in Figure 6.9), we can see that however much it is a powerful set of ideals, it is a firmly circumscribed construct. It is but one version of 'America', a version that leaves out great swathes of terrain and vast numbers of the population.

This is how discourses work; they naturalize what we talk about, what we think and how we behave and, when they are particularly powerful, they make it very difficult to imagine alternatives, or 'counter-discourses'.

O Beautiful,
For spacious skies
For amber waves of grain
For purple mountains, majesty
Above the fruited plains
America, America
God shed his grace on thee
And crown thy good,
With Brotherhood
From sea to shining sea!

Figure 6.9 America the beautiful – but what is left out?

4.2 A discourse of the environment

For many years, nascent environmental groups confronted a wider political discourse that altogether ignored environmental concerns. Environmentalists have struggled to gain wider acceptance for the idea that the Earth's resources are not infinite. The image of the blue planet as seen from the Apollo XI moon mission was an important breakthrough in environmental discourse: it drove home the concept of a finite Earth. This image was used to good effect by environmentalists to assert the belief that *all* Earth's resources are finite and that increased consumption is not in its or our long-term interest. Yet this understanding is in *contest* with other – particularly widely held liberal economic – discourses about limitless growth and the satisfactions of continuous consumption. There is a clear gap between environmentalists' belief in the 'reduce, re-use, recyle' mantra and the fashion for buying new, more and more often – a fashion that fuels what conventional liberal economists see as a healthy economy.

To achieve recognition for their ideas, environmentalists have tried a number of strategies, including media interventions, protests to government officials, local and multinational companies, and legal redress through the courts, whether at local, national or international level. In so doing, they have created new discourses about the environment and people's relationship to their environments.

The role of media interventions is discussed in Chapter Eight.

4.3 The environmental mainstream and the 'subaltern'

It would be simplistic to suggest that all environmentalists can be lumped into just one of two main groups: the mainstream and what Laura Pulido, a writer on environmental activism, calls the 'subaltern'. But, for our purposes here, the distinction is useful. According to Pulido, the dominant mainstream is characterized by access to large amounts of funding, and to political decision makers through expert political lobbying, a large and financially supportive membership and a 'preoccupation with [consumption and] quality of life versus subsistence, or production, issues' (Pulido, 1996, p.22). To put this in somewhat different terms, for mainstream environmentalists there has conventionally been a dividing line between the concerns of the environment and other social issue concerns. Pulido quotes a response to a survey of mainstream environmentalists: 'When I want wilderness preservation, I join a wilderness group. When I want civil rights, I support the NAACP' (Pulido, 1996, p.23).

The NAACP is the National Association for the Advancement of Colored People, a very important and, at times in its history, radical civil rights group in the USA.

Conventionally the mainstream environmental movement promoted a powerful discourse about the environment as *separate* from people: the environment defined as wilderness, woodlands, deserts and oceans, areas imagined as largely devoid of human habitation (see Chapter Two). In many instances, the public voice of the environmental movement encouraged the idea that protecting the environment means looking after particularly cuddly mammal or bird species and, sometimes, large scale fauna or landscapes. Other issues have been raised, such as a need to combat air pollution or struggles against the building of new roads, but advocates of these struggles have often been accused of middle-class NIMBYism (see, for example the discussion on wind farms in Chapter Three). What it did *not* do was focus on the concerns of rural farmworkers or worries about toxicity in urban areas. The concerns of mainstream environmental groups – such as the Sierra Club, World Wide Fund for Nature and, later, Friends of the Earth and Greenpeace – for a long time *tended* to emphasize the importance of non-human over human concerns. According to EJM advocates, where there was concern with the non-human world, it was largely focused on areas of leisure interest to wealthier and whiter groups of people. Why was this?

In the early 1980s a US study by Lester Milbrath (cited in Taylor, 2000) suggested that environmentalists were disproportionately (compared to the general population) male and over 20 years of age, more highly educated and had higher incomes. The sample surveyed in 1980 was also disproportionately white. At the time, census data indicated that white peoples constituted 83.2 per cent of the US population and black peoples 11.7 per cent. Within the environmentalist sample, however, whites constituted 92 per cent and blacks a mere 2 per cent. By 1982 the percentage of whites had *increased* to 94 per cent (see Taylor, 2000, p.539).

Similar patterns continued to exist in US mainstream environmental organizations throughout the 1980s and 1990s. Doretta Taylor cites statistics from a 1992 study, which demonstrate the extent to which staff, volunteers and even membership of 63 mainstream environmental organizations were virtually exclusive of non-whites (cited in Taylor, 2000, p.539).

One reason perhaps why dominant environmental discourses have tended to sound middle class and why the concerns of non-white communities have tended to be left off the agenda of mainstream environmental groups is the make-up of these large, influential groups (see Figure 6.10). In recent years, livelihood issues, particularly livelihoods of indigenous peoples, have figured more prominently in mainstream campaigns. What this suggests is that there was a material reason for the emergence of the conventional, mainstream environmental discourse. Importantly, there are also material effects of the tendency to ignore the environmental interests of non-white, non-middle class groups.

Figure 6.10 Workers in mainstream environmental organizations initially tended to be predominantly white; shown here is a 1975 meeting of Greenpeace's 'Save the Whale' campaigners.

This tendency and these material effects have spurred the creation of 'subaltern' environmental groups with radically different agendas. The priorities of the EJM – which Pulido would categorize as emerging from the subaltern – and their understanding of the environment is in many ways at odds with that of mainstream environmentalists. Interestingly, unlike the mainstream environmental groups of the 1980s and 1990s, subaltern groups tend to have a far greater percentage of supporters from women and from non-white and working class groups, so one could argue there is a material interest behind the emergence of their subaltern discourse.

For the EJM, the apparent dividing line between the natural and the human is tenuous at best and the environment is understood less in terms of wilderness that is 'out there' or separate from people, but instead is seen, very simply, as *where we live*. This conception of the environment as having proximal immediacy is an important aspect of the story of Warren County, North Carolina, and the protests of that community against toxic dumping. It is with this story that the next section begins.

Summary

Broadly speaking, our understanding of the environment is contested terrain:

- The mainstream discourses that inform our ideas about the environment are powerful in large measure because they seem so 'natural'.
- Nevertheless, material experiences may prompt people to reconsider their understanding of the environment and to pose alternative discourses. This is an important part of the political struggle over the environment.

5   Power and the environmental justice movement – the case of Warren County

In 1978 three men, hired by the Ward Transfer Company, drove liquid tanker trucks containing polychlorinated biphenyl (PCB) contaminated oil along 210 miles of state roads in North Carolina, intentionally allowing 30,000 gallons to drain out along the roadside (Bullard, 2000, p.30). For two weeks their activities went on unabated, creating an environmental disaster and a massive public health problem (McGurty, 2000, p.376). Once the roadside pollution was established, however, the state government was left with a major political dilemma: it needed to clean up the roads, but where to put the now highly toxic soil? An apparent solution presented itself when a bankrupt farmer agreed to sell his land in Afton, Warren County, and the state proposed to use the area for landfill (McGurty, 2000, p.376). Afton was a mainly rural area, and its inhabitants predominantly working class and black. Three years of legal action followed the state's announcement, with nearby residents contesting the landfill on the basis that the site was geologically unfit. In 1982 the residents lost their battle and state workers began to prepare the site. Residents, however, were not ready to accept the court decision (Pinderhughes, 1996, p.240).

5.1   Civil rights protest and the beginning of a new environmental movement

A coalition of local landowners and civil rights activists began a campaign of civil disobedience (marches, demonstrations, roadblocks, etc.) to stop work at the site and to deny access to the trucks transporting the toxic soil (see Figure 6.11). Ultimately the coalition was unsuccessful and the landfill was established. There is, however, an enduring legacy of this episode, and that is the broader development of the EJM.

Figure 6.11 Over 400 people were arrested at the protests in 1982 against the Afton dump: (left) protesters attempt to block a highway to deny the trucks access, and (right) hundreds of protesters marching, North Carolina.

Prior to the mid-1980s and the Love Canal and Warren County fights, US environmentalists were, as we have seen, readily characterized as predominantly white, male, middle class and more inclined to be concerned about non-human habitats. The events in Warren County marked the beginning of a coalition that brought together diverse constituencies, including these more conventional environmentalists, and civil rights activists and local residents not known for their political activism – thus radically changing the face of environmentalism.

Together, these groups challenged orthodox environmental discourse, suggesting that the environmental problems assailing Warren County were also political problems – problems of environmental injustice. Activists asserted that toxic waste, its management and disposal, were deeply political issues closely tied to race and class. Making these links meant that what might have been a single-issue campaign about toxic or hazardous waste was transformed into a wider, political movement for environmental justice. As a political movement, concerned with the everyday lives of people and their environments, the EJM was able to connect its concerns to a range of social issues. It became the EJM's mission to ensure change; change at the material level and change in terms of discourse. The EJM sought redress for the victims of toxic waste disposal in the form of housing reallocation or compensation (distributive justice). However, the movement also sought to change the very terms of reference for environmental considerations to include *a priori* consideration of the impact of environmental decisions on human populations (procedural justice). Moreover, it sought to challenge environmentalist orthodoxy, to ensure that, for instance, cityscapes and even rural 'wastelands' fell within the ambit of environmental concern.

5.2 Environmental racism and the expression of environmental inequality

In the preceding sections, we looked at the way some groups – those perceived as politically disenfranchised or in some way peripheral – have struggled to challenge the power of industry and government. In so doing, we considered the

EJM proposal that environmental inequalities occur *because of* already existing inequalities: poorer people, those in rural or poorer urban communities, are more likely to suffer greater problems with environmental pollution than those in wealthier communities. But for some writers and activists, there is an important dimension to the equation that links social or political and environmental inequalities in what is called 'environmental racism'. Environmental racism is said to be evident, for instance, when a disproportionate number of toxic land-fill sites are located in black neighbourhoods compared to white neighbourhoods, *despite* there being no technical, geological or safety reasons for this.

According to the EJM, it is important to recognize the different ways in which environmental injustice, including environmental racism, is expressed. They identify the following dimensions:

- *Location, or proximity*
  For lower income groups there is a greater likelihood of greater proximity to a greater number of environmental 'bads'. For higher income groups there is a greater likelihood of greater proximity to a greater number of environmental 'goods'.
- *Access to opinion makers and to the media*
  Where there are environmental problems, public concern is generated unevenly depending on the area at risk (though note that some types of risk, e.g. nuclear hazard, more readily attract media attention). Poor rural areas, industrialized areas and areas already suffering from some form of environmental degradation are less likely to make the headlines or, if they do, publicity has little effect, as the constituency is of marginal interest to political policy makers. Pristine areas of 'outstanding natural beauty', areas to which the wealthy tend to have prime access, are less likely to suffer environmental atrocities, but if they do, they also make the news.
- *Rhetoric*
  For years, environmental discourse was defined by middle class whites who were the prime constituency of the mainstream environmental movement. 'The environment' was constituted in legislation, in popular discussion and in song by groups with seemingly remote concern for the wider human implications of environmental use and misuse. With 'the environment' defined as areas of non-human habitation, the city environment or the already polluted environment were outside the prevailing environmental discourse.
- *Government intervention*
  In the USA, environmental risks have been distributed along paths of least resistance. In the past, this has meant sending toxic waste to landfills in predominantly poor, black rural communities or onto Native American land (e.g. the contested proposal to establish a nuclear waste interim storage site on the Goshute Native American Indian Reservation in Colorado). Where environmental racism has become untenable on its own soil, environmental hazards created in the USA and other developed nations have been shipped 'elsewhere' for disposal (for a discussion on this practice and resistance to it,

see Blowers, 1996). This has recreated the very practices and policies that resulted in the EJM in the USA, but has globalized them.

So why have environmental injustices occurred? We have already gone some way to answering this: mainstream environmental discourses traditionally precluded consideration of human habitats; areas perceived to be degraded have been seen as preferred sites for *further* degradation; and people in poor areas have less political voice and – until the EJM emerged – little vocabulary with which to contest LULU decisions.

5.3 The seminal tale of Warren County

According to Bullard, Warren County is a seminal tale of power and environmental justice. Some economic statistics set out the story:

> The Afton community [in Warren] is more than 84 percent black. Warren County has the highest percentage of blacks in the state and is one of the poorest counties in North Carolina. [In the 1980, blacks] ... composed 63.7 percent of the county population and 24.2 percent of the state population. Per capita income for Warren County residents was $6,984 in 1982 compared with $9,283 for the state. Although the county lags far behind the rest of the state on a number of economic indicators, over three-fourths of Warren County residents own their own homes. Why was Warren County selected as the PCB landfill site? The decision makes more political than economic sense.
>
> (Bullard, 2000, p.30)

The main geological problem with the Afton site was its high water-table. This, combined with the fact that most of the community's residents took their drinking water from local wells, meant that the likelihood of PCBs leaching into the groundwater carried potentially extremely dangerous risks to human health (Bullard, 2000, p.31). Still, the state government went ahead and sited the landfill in Afton. Despite meeting with hundreds of protesters (in the space of two weeks 414 protesters were arrested) the trucks kept coming, dropping over 6,000 truckloads of PCB-contaminated soil in Warren County. According to Bullard and others, the choice for the site centred on a few key political 'truths':

- black communities such as in Warren County tend to have neither the economic nor the political clout to fend off environmentally destructive activities, nor the financial wherewithal to move away from polluted areas;
- industry and politicians follow a 'path of least resistance' siting where there are likely to be least complaints, less media interest and fewer (or lower) pay outs should legal action occur;
- as a consequence of the above two points, the right to live in a healthy environment has more often been denied to black and lower income groups than to white, middle and upper income groups.

One further point: blacks and poorer whites tend to live in areas that are already environmentally compromised. As we saw with the peripheral communities in Chapter Three, that pre-existing factor often leads to further degradation. Bullard argues that whether we look at the lives of 'Latino Americans in the Southwest or native Americans in the West, people of colour bear a disproportionate share of the nation's environmental problems' (Bullard, 2000, p.xvi). In short, he claims, this adds up to a policy of environmental racism.

Activity 6.3

If poor, non-white areas carry the lion's share of environmental risk, including intentional dumping of toxic wastes, does this necessarily add up to a *policy* of environmental racism? What do you think? Try to identify some reasons why it might *not*. (You might also like to refer back to Chapter Three for ideas on this point.)

Comment

- There is some evidence to suggest that Bullard overstates the environmental racism position. Love Canal is a good example of a predominantly white community suffering the same kind of environmental degradation that Bullard suggests is primarily visited on non-white communities. Perhaps, then, thinking about power as resource, this is a problem of access to political resources as much as it is an issue of racial discrimination.
- 'Policy' suggests conscious legislation. Arguably, even when decisions are made that are consistently to the disadvantage of one group over another, this *may* not be deliberate. Thinking about the idea of power as discourse, we might conclude that the very idea of managing toxic wastes away from the areas that have been used conventionally could have become *unthinkable*: it literally falls outside the realm of possible thought. This essentially but not intentionally orders policy, as an effect of discourse.
- We might also consider a more cynical position: in a culture where civil rights are such an important part of the political debate, there are obvious advantages to linking one's cause to issues of racism (or gender equality, for that matter). Identifying actions as policy rather than incidental or arbitrary lends weight to the argument.

5.4 Creating a new discourse: environmental justice as civil right

The struggle against environmental racism has proved an important focus for the EJM. Using the rhetoric of civil rights, US environmental justice activists have fought for legislation to include the right to a safe and healthy environment along with other human and legal rights, such as the right to equal treatment before the law, to education and to non-discriminatory practices in the workplace, in health treatment, housing, schools, etc. In a culture where issues of racism and political correctness have become a regular part of public debate, the weighty chant of 'environmental racism' has helped to heighten the profile of environmental justice. We could say that the competing, subaltern – as opposed to dominant – discourse of environmental justice has added a strong voice to the civil rights chorus. And, as shown above, mainstream environmentalists are now also chiming in – even regulators and legislators regularly hum a few bars of environmental justice.

The rhetoric of environmental justice has made its way into environmental legislation in the USA, thus encouraging many groups to fight their cause with painstaking reference to 'environmental justice'. It is important to recognize, however, that environmental justice campaigns – and the compelling rhetoric of 'environmental justice' – are not located solely in the USA. Not surprisingly, environmental and social justice issues are actively linked in many parts of the world, in urban and rural areas, in small villages and large cities, and in the developing as well as developed worlds.

In this section, I have intimated that the relationship between values, power and action is tightly woven. Where there are competing values, struggles occur. People take action to alter power relations and part of that action involves challenging the values that underpin differences in power. It is not enough, according to EJM activists, to gain reparation for toxic dumping; it is essential to challenge the very beliefs and attitudes that *support* the policies that enable such practices to occur. In short: the material and the discursive are very closely linked and it is little wonder that a struggle over one involves a struggle over the other. In the following section, I will expand somewhat the initial concept of environmental justice. We take as our starting point the argument that environmental atrocities are visited not only on the poor and minority groups in the USA, but are closely connected with the unequal distribution of power across a range of social differences. I will develop this argument by taking into account one additional factor: globalization.

Summary

- The case study of Warren County helped bring to light the issue of environmental racism. The EJM has drawn very much on the approach of the civil rights movement with, for instance, staged protests and a refusal to accept unjust legal decisions and policies.
- The notion of discourse needs to be acknowledged; EJM activists have sought to counter traditional mainstream environmental discourses.
- The material and the discursive are closely interwoven: struggle around one perhaps inevitably involves struggle around the other.

6 The export of tragedy: environmental injustice and the case of Ogoniland

The story of Ken Saro-Wiwa is one of many stories about how a global economy impacts on local communities and their environments. It is also another story of environmental (in)justice.

6.1 Ken Saro-Wiwa and the battle for Ogoniland

Before his death on 10 November 1995, Ken Saro-Wiwa brought to international attention the exploitation of Ogoni land and people wrought by multinational oil companies working hand-in-glove with the Nigerian military dictatorship, led by General Sani Abacha. Despite the world's attention, however, the Nigerian authorities resisted all international political pressure, and on the grounds of treason, Sara-Wiwa (Figure 6.12) and his compatriots were executed.

Figure 6.12 Ken Saro-Wiwa, Ogoni leader and internationally known political journalist. Saro-Wiwa was hanged for political and economic treason; his crime was to demonstrate against his government's human rights abuses and the environmental degradation of Ogoni land by oil companies Royal/Dutch Shell and Chevron. He is seen here at the Ogoni Day demonstration, 21 October 1993.

According to Francis Adeola, a political researcher, Ogoni lands and the Ogoni people suffered greatly for many years before Saro-Wiwa managed to bring their devastation to light. Adeola explains that Ogoniland, in particular its many oilfields (see Figure 6.13), has been a major contributor to the national economy. Despite this, there has been little in the way of environmental interest taken in the area and little in the way of protection or economic gains for the Ogoni people.

In addition, the area is home to a number of polluting industries, including oil refineries, a major fertilizer plant and a large petrochemical plant (Adeola, 2000, p.695).

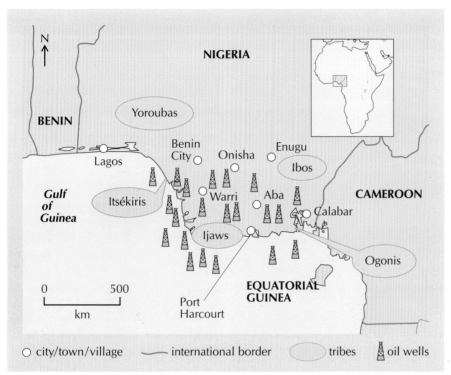

Figure 6.13 Location of Ogoniland, near Port Harcourt in Nigeria, and cluster of oil sites.

Reports by the Sierra Club suggest that:

> Royal Dutch Shell and other associated-[multinational oil corporations] (e.g. Chevron Corporation) ... have taken more than $30 billion from Ogoniland, leaving behind ecological devastation, destitution, environmentally-induced illnesses, and a shorter life expectancy among the people.
>
> (Adeola, 2000, p.695)

The environmental plight of the area (Figure 6.14) is confirmed by the devastation of the native mangrove trees. According to the US Energy Information Administration, the mangrove, which once served as both fuel and habitat, is now 'unable to survive the oil toxicity of its habitat' (US Department of the Environment, 2000).

Years of neglect, particularly during the period of military rule, led to the subsequent, civil government instigating more stringent agreements for acceptable environmental practice with the various multinational oil companies. President Olusegun Obesanjo made public his concern about the 'poor performance' of the oil multinationals which, he believed, failed to meet

The Sierra Club is an organization which for many years took a deeply conservative mainstream approach to the environment, concerning itself primarily with wilderness conservation. In recent years, however, it has revived its earlier more radical stance and added its voice to a number of environmental justice campaigns.

Figure 6.14 The legacy of exploitation in the search for oil: (a) Shell pipeline runs through the centre of Kenkana village; (b) oil spill flaring; (c) landscape devastated by oil spill; and (d) an Ogoni man collecting water from an oil polluted stream, Bori region.

international standards. Interestingly, in June 2000, a Nigerian court found against Shell in a dispute over a large oil leak in the 1970s in Ogoniland. The court ordered Shell to pay US$40 million compensation for the damage. Shell contested the ruling.

In another incident, in 1973, one of Shell's pipelines burst, spilling an estimated 2,000 barrels of crude oil, affecting local land and the Osadegha stream, which runs through farmlands and feeds some of the important water sources in the area. Shell initially denied responsibility for the spill, and then suggested the spill was caused by sabotage. Subsequently, however, Shell agreed a moderate pay-out in local currency (Sierra Club, 1999). In the 40 years to 2001 more than 4,000 oil spills were recorded in Nigeria's Niger Delta (US Department of the Environment, 2002).

There is little doubt that the area which comprises Ogoniland has suffered serious environmental damage. This brings us back to our first key question, in this case to ask: how and why has the contest over Ogoniland arisen? An answer to this lies, according to Saro-Wiwa, in political inequality in Nigeria more widely.

6.2 Minorities, majorities, power and protest

According to Adeola and others, a key reason why Ogoni land has been despoiled and Ogoni people have not been adequately compensated is that the Ogoni are a minority population in Nigeria and in the immediate Niger Delta area. This has contributed to the Ogoni having had fewer opportunities to gain access to political structures at either a local or national level. In the Ogoni Bill of Rights (publicly presented in 1990), the signatories claim that at that time, Ogoni people had 'No representation whatsoever in ALL institutions of the Federal Government of Nigeria'. Moreover, they had 'no pipe-borne water; no electricity; no job opportunities for the citizens in Federal, State, public sector or private sector companies' (Ogoni Bill of Rights). The Ogoni's relative political weakness – as a minority group – arguably made exploitation by the government-sanctioned, multinational oil companies all the easier.

Put simply, with fewer people, minorities have fewer claims to power. Moreover, they are easier to control in a physical sense: because of their smaller numbers any protest is more easily quelled or crushed. In this model of power, the bottom line is the manifest physicality of power. Despite the natural resources of their land, the Ogoni had few other resources to draw upon, whether political or social, because of their historical marginalization. It is worth noting that marginalization of minority interests occurs in both democracies and dictatorships. To explain this, we need to remember the idea of power as discourse. For if Ogoni ideas or beliefs are treated ordinarily as outside the mainstream, as troublesome or against majority interests, if the Ogoni are, over a long period of time, 'discredited', it may become 'natural' to treat not only their ideas but the people themselves with derision and disregard. Similarly to Bullard's notion of environmental racism, we might suggest, there was a culturally ingrained tendency in Nigerian politics to discrimination that operated against Ogoni interests.

So, the Ogoni being exploited by the oil companies in Ogoniland was not the whole story – there existed social and political conditions within Nigerian politics that meant the Ogoni were both resource poor and discursively marginalized.

6.3 The bases of protest

From an environmental justice perspective the extent of this marginalization provides ample evidence of the interconnectedness of environmental concerns with other social and political concerns. For the Ogoni, in fact, the question of environmental justice is fundamentally a question of survival.

It is fascinating to see how the different kinds of protest being made by the Ogoni have come together and been articulated in a discourse of environmental justice.

The Ogoni have protested against:

- Shell and other oil companies' incursion onto Ogoni land;
- the Abacha government's complicity in the environmental degradation resulting from oil extraction;
- lack of compensation for despoliation;
- the problems the minority Ogoni have traditionally had in gaining political rights in Nigeria;
- globalization (in terms of trade and economic 'imperatives' as touted by multinational corporations).

From this list, you can see how discourses work to articulate together different concerns. In this case, while a range of issues is tied to the Ogoni's complaint against Royal Dutch Shell, the 'single issue' is linked, discursively, to political, economic and environmental concerns which can be summarized as conflict over 'environmental justice'.

6.4 Global networks: Ogoni protest and international action

Traditionally, the Ogoni have been a vocal minority. But recognizing that political protest at local or national level would continue to have little effect, leaders such as Saro-Wiwa took their complaints to authorities outside Nigeria. In 1992, the Ogoni's case was heard by the United Nations Commission on Human Rights, and in 1993 the Ogoni joined as a member of the Unrepresented Nations and Peoples Organization. Subsequently, a number of international environmental and human-rights NGOs launched campaigns to raise the profile of the Ogoni's plight. In the weeks preceding Saro-Wiwa's execution, Greenpeace, the Sierra Club, Amnesty International and a coalition of other NGOs promoted high-profile boycotts of Shell Oil, with protesters blockading Shell Oil petrol stations (Figure 6.15), erecting banners outside the corporation's head offices and calling on the international community to intervene against the military tribunal decision. There was a moment in the midst of this ferocious outcry when it seemed impossible for the Nigerian authorities to go ahead with the execution, and yet it happened.

In the wake of Saro-Wiwa's execution and the surrounding international outcry, Shell has paid some compensation and some attempts at clean-ups have been made. Obesanjo's government has indicated that the oil companies' poor environmental behaviour would no longer be tolerated and there is some indication that the resident oil companies are complying with the new government's directives.

What we might conclude from this case study is that while globalization may contribute to environmental degradation in some parts of the world, levels of globalized communication may be of benefit to marginalized communities throughout the world. For the Ogoni, being able to articulate the discrimination of the Nigerian government in terms of environmental injustice meant that they were able to garner support not only from groups such as Amnesty International, who were concerned with human rights abuses, but also from groups such as Greenpeace, who saw their problems as environmental issues.

In terms of the development of the environmental justice movement, the Ogoni discourse was instrumental in bringing together such NGOs in a recognition of the inextricable relationship between environment and livelihood. The articulation of the Ogoni's local plight with wider issues of globalization struck a chord with a growing international movement against multinational and transnational corporations (see Chapter Five), which focused on difficulties in holding such companies accountable for their actions.

It is this linking of social, environmental, political and economic causes that has provided a great boost to environmental justice at local, national and international levels.

Figure 6.15 Greenpeace protest at a Shell petrol station.

Summary

- The story of environmental justice in Ogoniland is an example of globalization and the unequal distribution of power. The Ogoni land and people, rich in oil but poor in resources to combat the power of transnational oil companies and the Nigerian authorities, were exploited in the acquisition of oil.

- The passionate message articulated by the Ogoni, particularly Ken Saro-Wiwa and his compatriots, fitted with that of international activists in the struggle for political, economic and environmental justice. The complementary discourses made them natural allies and strengthened the protests for all.

7 Conclusion

The three case studies illustrate how civil rights activists have come together with environmentalists to overturn social and environmental injustice in practice and to change the prevailing discourses and associated power relations. Thinking about both the similarities and differences between our three case studies raises some important considerations.

The first is to do with the 'translation' of concepts/ideas from one country or one culture to another. In the USA, the linking of environmental injustice to social injustice or racial discrimination marked a major change in the way 'the environment' and ideas about the poor and racial and ethnic minority discrimination were understood. However, we must not assume that the same thing happened in Nigeria. In the USA, the EJM formed, in part, as a reaction against the *mainstream* environmental movement and its segregation of wilderness concerns from concerns with people. Nigerian environmental history is different. In fact, environmental issues in Nigeria were not decoupled from social issues in the same way as in US mainstream environmental discourses. In other words, while 'environmental justice' was a new concept for many in the West, the premise of environmental justice was rather conventional for the Ogoni (and for many developing nations).

The second consideration is about differences in circumstance between the inhabitants of Ogonilands and those of Love Canal and Warren County. As noted above, many residents of Love Canal were tied through employment to the very industries responsible for polluting their living environment. This made it difficult for homeowners to initiate complaints against these companies. This was a less significant issue for Ogoni people as very few were employed by Shell. A prime issue here is that the Ogoni got little back from oil extraction and consequent despoliation of their land.

In the case of Warren County, there are other differences. While Ogoni land in the North East of the Niger Delta was one of great riches, with large oil reserves, farmland and mangrove woods, the largely rural Warren County was relatively less productive. What is similar about Ogoniland and Warren County is that the marginalized groups concerned received little or no compensation for the degradation of their lands: local people were not employed by the polluting industries, nor did they reap financial rewards in other compensatory arrangements (such as local amenities, or pay-outs for pollutant clean-ups).

Interestingly, at Love Canal, the predominantly white residents found it virtually inconceivable that their employers, the local government and the authorities more generally would subject them and their families to toxic waste. For the Ogoni and Warren County populations, this kind of action was still abhorrent, but perhaps less of a shock. In both latter cases, long-term discrimination, evidently on the basis of ethnicity or race, meant that these groups found environmental abuses fit into a set of discourses already at their disposal. In the case of Warren County, the decision to site the PCB dump in an evidently geologically untenable location was quickly identified as another discriminatory action by racist authorities, operating in a political and economic climate that prioritized wealthier, white neighbourhoods. What was new about the way the Warren County protests developed is that social injustice and institutionalized racism were articulated in terms of environmental concern. Given that this was not a dominant framework of understanding for either civil rights activists or environmentalists this marked an important change in approach to toxics campaigns for both.

For the Ogoni, exploitation of land they worked and lived upon was historical and premised on the difficulty with which minority groups might protest. In the Ogoni case, the national government benefited financially from turning a blind eye to the polluting practices of Shell and other oil companies. At the extreme, this led to military assistance from the government when Shell complained about protest blockades near their oil processing plants. Although evidence is not entirely clear on this point, Amnesty International asserts that Shell has admitted inviting the Nigerian army to its operations on Ogoni land, and 'provided them with ammunition, logistical and financial support for a military operation' (Amnesty International USA, 2000). It is important to recognize such power relations as real conflicts of interest, sometimes held at bay through discourse and sometimes breaking loose into more physical action and conflict.

For Shell, what they saw as legitimate business agreements with the Nigerian government were challenged and materially put at risk by minority Ogoni protest. For the Ogoni, physical demonstrations were not enough; a clear counter-discourse that challenged the powerful interests of the government and Shell was necessary to ensure their protest reached organizations likely to have the political will and influence to be able to help them. In this case the Ogoni and their supporters put forward a 'David and Goliath' narrative which pitted the small, local Ogoni against the Nigerian dictatorship and the multinational oil companies. The Ogoni made claims for their land that distinctly tied their political freedom to environmental self-determination in similar ways to the US EJM, but added to this was a struggle against economic globalization – or global trade of the kind demanded by multinational corporations such as Shell and Chevron (see **Castree, 2003**).

In all three cases residents' subaltern positions meant that protest was difficult, but not impossible. Despite their shortcomings in terms of access to resources, the Love Canal and Warren County residents and Ogoni people *were* able to act – to bring their cause to a wider public and, through media exposure, garner sympathy with other, more powerful groups, through court battles or international pressure groups – to achieve some measure of environmental justice.

We have seen, therefore, that the EJM was born out of particular political, economic and environmental conditions in the USA. It is premised on the notion that discrimination exists between different groups, which results in the economically or politically privileged groups benefiting environmentally at the expense of less privileged ones. According to Bullard, the dividing line between these two groups is primarily accounted for by race, but both he and other commentators have argued that discrimination along class and gender lines is also significant. The notion of environmental justice, initially part of a critique of the mainstream environmental movement, has now been adopted by many mainstream groups as an important conceptual tool in campaigns to encourage multinational and governmental environmental responsibility. In each of the three case studies, we saw different reasons for the campaigns' advocates seeking environmental justice, but nevertheless the importance of the term as a rallying cry.

We also looked at the notion of 'discourse' in order to understand how power works not just in terms of 'power over' (for example physical power), but also in terms of everyday practices that become naturalized or common sense. This revealed how relatively easy it was for authorities, institutions and majority groups to ignore the concerns of minorities with regard to the environment and their everyday experiences. Environmental justice is part of the reconceptualization of environment that *includes* rather than *precludes* the interests of people. As our attitudes toward our environment change, our environment – or at least what we count as our environment – changes too. Of course, this is much more than just a question of perception – for different ideas lead to different actions. Although discourses are about ideas and attitudes they have 'real world' effects. As you will see in Chapter Eight, the battle for the environment is very much a battle for hearts and minds, and the notion of discourse is a helpful tool in understanding the ground on which that battle takes place. The concept of discourse helps us to make sense of how dominant ideas come to have such significant effects on our lives, socially, culturally, politically and materially – *and* how subaltern views can eventually emerge to challenge them.

For the environmental movement as a whole, a discursive shift to environmental justice means that the concerns of environmental NGOs can now include the plight of, for instance, forest-dwellers and city-folk as well as issues relating to photogenic mammals and wilderness access. The EJM has ensured that the contest over the environment and 'environmental justice' is recognized as a political issue – one that raises the question of wider inequalities and is connected to the uneven exercise of power by government, industry and a range of other interest groups. The grass-roots activism of environmental justice campaigners has emerged as an important means of challenging the authority and ideas of the mainstream – both within the more orthodox environmental movement and society at large at local, national and international levels. Moreover, the conceptual link between the environment and notions of rights is leading to legislation, at both national and international levels, which seeks to protect this relationship between people and their environments.

References

Adeola, F. (2000) 'Cross-national environmental injustice and human rights issues: a review of evidence in the developing world', *American Behavioral Scientist*, vol.43, no.4, January, pp.686–707.

Amnesty International USA (2000) 'Just Earth: Nigeria', www.amnestyusa.org/justearth/countries/nigeria.html

Blowers, A.T.(1996) 'Transboundary transfers of hazardous and radioactive wastes' in Sloep, P. and Blowers, A.T. (eds) *Environmental Policy in an International Context*. London, Arnold with Open Universiteit, The Netherlands and The Open University.

Brandon, M. and Smith, S. G. (2003) 'Water' in Morris et al. (eds).

Bullard, R. (2000) *Dumping in Dixie: Race, Class and Environmental Quality* (third edition) Colorado, Westview Press.

Castree, N. (2003) 'Uneven development, globalization and environmental change' in Morris et al. (eds).

Cock, J. and Fig, D. (2000) 'From colonial to community based conservation: environmental justice and the national parks of South Africa', *Society in Transition*, vol.31, no.1, pp 22–36.

Dickens, C. (1854) *Hard Times*, London, Chapman and Hall Ltd.

Dobson, A. (1988) *Justice and the Environment: Conceptions of Environmental Sustainability and Dimensions of Social Justice*, Oxford, Oxford University Press.

Earthday Ecological Footprint, http://www.earthday.net/footprint/index.asp (accessed on 7.8.02).

Edwards, B. (1995) 'With liberty and environmental justice for all: The emergence and challenge of grassroots environmentalism in the United States' in Taylor, B.R. (ed.) *Ecological Resistance Movements: The Global Emergence of Radical and Popular Environmentalism*, New York, State University of New York Press.

Hinchliffe, S.J. and Belshaw, C.D. (2003) 'Who cares? Values, power and action in environmental contests' in Hinchliffe et al. (eds).

Hinchliffe, S.J., Blowers, A.T. and Freeland, J.R. (eds) (2003) *Understanding Environmental Issues*, Chichester, John Wiley & Sons/The Open University (Book 1 in this series).

International Wildlife (1994) 'Clinton orders environmental justice for all', vol.24, no.3, May/June, p.26.

McGurty, E. M. (2000) 'Warren County, NC, and the emergence of the EJM: unlikely coalitions and shared meanings in local collective action', *Society and Natural Resources*, vol.13, no.4, pp.373–88.

Morris, R.M., Freeland, J.R., Hinchliffe, S.J. and Smith, S.G. (eds) (2003) *Changing Environments*, Chichester, John Wiley & Sons/The Open University (Book 2 in this series).

Pinderhughes, R. (1996) 'The impact of race on environmental quality: an empirical and theoretical discussion', *Sociological Perspectives*, vol.39, no.2, pp.231–49.

Pulido, L. (1996) *Environmentalism and Economic Justice: Two Chicano Struggles in the Southwest*, Tucson, University of Arizona Press.

Rothman, H. (1998) *The Greening of a Nation? Environmentalism in the United States since 1945,* Fort Worth, Texas, Harcourt Brace.

Sierra Club (1999) 'ERA Monitor report no.8: six year old spillage in Botem-Tai', www.sierraclub.com/human-rights/nigeria/background/spill.asp

Szasz, A. (1994) *EcoPopulism: Toxic Waste and the Movement for Environmental Justice*, Minneapolis, University of Minnesota Press.

Taylor, D. (2000) 'The rise of the environmental justice paradigm: injustice framing and the social construction of environmental discourses', *American Behavioural Scientist*, vol.43, no.4, January, pp.508–81.

US Department of the Environment, Energy Information Administration (2002) January, www.eia.doe/emeu/cabs/nigeria2.html

US Department of the Environment, Energy Information Administration (2000) April, www.eia.doe.gov.emeu

Chapter Seven

Environmental values in environmental decision making

Jacquie Burgess

Contents

1 Introduction

Imagine the scene. You're sitting in a timber-beamed room above the bar in a Sussex pub. It's a rather cold, blustery March evening with the wind whipping across the marshes from the sea. Seated around a small table are eight men and women who have got to know one another quite well over the last three weeks. All they had in common at the start of their acquaintance was their agreement to answer questions put to them by a woman who knocked at the door one evening and said she was carrying out a survey for English Nature. The group are discussing the questionnaire they had answered that night, and how they tackled the question asking them how much they would be willing to pay for a wildlife enhancement scheme on the Levels.

Jacquie	But can you say that a sum of money is a way of expressing what the Marshes are worth to you?
Greg	It's not a very good measurement but it's probably the only way we've got. Because – 'For a thousand pounds, would you be prepared to see the Marshes go?' The answer's 'no!'
Meg	The trouble is – I think that if you feel passionately about something, it's difficult to put a monetary value on it. And I think perhaps somebody who doesn't feel impassioned about the Levels and its importance, perhaps they're the only people who could think seriously about putting a monetary value on it. Because to most people who care passionately about something – like most passion it's beyond monetary value. It goes onto a different level, doesn't it?
Malcolm	Well, you're talking about our very existence, really. If you're talking about the environment, then there isn't a price because it is our life, really. Our future life.
Christopher	That's what I meant earlier. That there is somebody who isn't looking at their 'pride and joy'. And they're looking nationally. And it's not a financial thing. But they're looking, relatively, at how important things are. And saying: 'yes, we can give away this little patch here'. And it's only when that little patch starts to encroach into what's important to you, that it becomes your local problem.
Carol	It's a much bigger issue than all this, isn't it really? Now everything has to have a price on it, so we're told. I mean, I think it's rubbish! But I'll be quite honest. When I was asked that money question, because of the financial situation my husband and myself are in, my answer was 'No, I couldn't afford to give anything extra'. At the moment we're stretched to our limits because of losing jobs and that sort of thing.
Meg	But the trouble is – that could be misinterpreted, couldn't it?

Carol	Absolutely. That's right.
Meg	As if you don't want it, you don't care.
Carol	It's a pity that everything has to have a monetary value. It's just our whole culture at the moment. And it needs people to stand up and say 'we cannot put a price on this. It's priceless – but that's no reason for building on it'.
Christopher	But you could have even greater problems, potentially. Suppose everybody who was interviewed was so passionate about this area. And everyone said they'd give too much. When they came to ask for it and people didn't have it to give – then what would happen? [pause, laughter].
Carol	That comes back to answering this honestly, doesn't it?
Meg	Not pretending you're some great giver, just to impress the interviewer!

The people whose thoughts and feelings you have just shared live around the Pevensey Levels, wet pastureland close to the sea between Lewes and Hastings on the south coast of England. As you can see from the map (Figure 7.1), most of the Levels' 4000 hectares is a Site of Special Scientific Interest and protected under EU legislation because of the rare plants and insects found in the drainage ditches: Pevensey is one of only two sites in the UK where the fen raft spider is found, for example. In addition, large populations of birds including lapwing and golden plover rest here during the winter.

fen raft spider
(*Dolomedes plantarius*)

Unlike other landscapes, relatively few people recognize the aesthetic or the ecological values of flat, wet, open land (see Figure 7.2). This makes it difficult to mobilize widespread or effective opposition to fight development proposals concerning such areas. And yet, for a minority, the beauty of wide skies, grey-blue water and green land is matchless. No other landscape can call up the same passion within those individuals – and I must confess that I am one – who love marshes. However, the frustration and anger we feel when our beloved landscapes fall under plough or concrete is not widely shared. The problem seems to be that other groups in society do not value wetlands to the same extent, or in the same way.

How we value our environments is discussed in further detail in Chapter Two.

lapwing (Vanellus vanellus)

The issues I want to explore in this chapter relate to our central theme of contest and have already been raised by the people in the Pevensey group discussion:

- How can the different values that people attribute to landscapes, nature and biodiversity be accommodated fairly in decisions about alternative courses of action?
- How much power do local people have to influence these decisions?

In particular, I want to explore one of the most common decision-making tools that economists and policy makers use to help resolve value conflicts over future uses of land and natural resources. I shall introduce you to the principles of cost–benefit analysis (CBA) and we will explore one important assumption of the

golden plover (Pluvialis dominica)

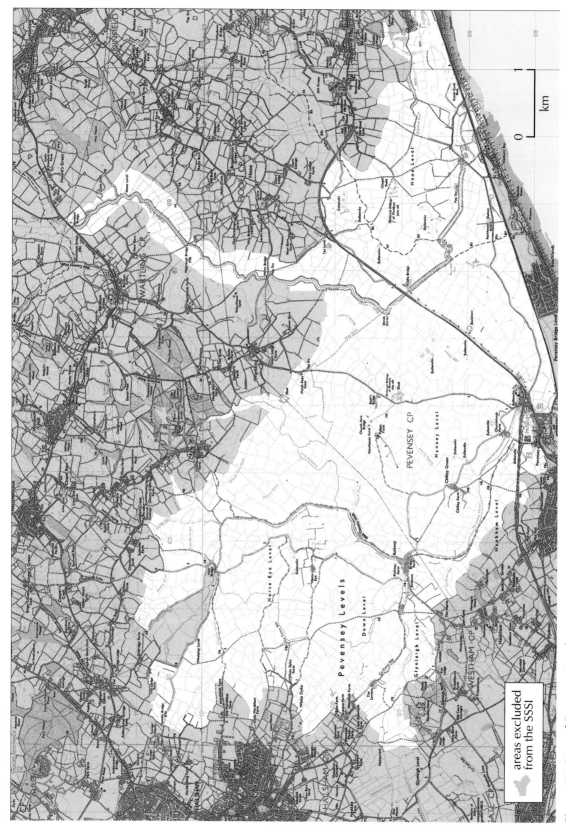

areas excluded
from the SSSI

Figure 7.1 Map of Pevensey Levels

Figure 7.2 The Pevensey Levels, as the name suggests, consists of flat, open wetland. Few people can immediately see the aesthetic or ecological value of such a landscape.

approach: that both instrumental values (roughly equivalent to use values) and non-instrumental values (roughly equivalent to non-use values) for nature and the environment can be expressed through sums of money. We shall also discuss how members of the public are engaged in monetary valuations through the use of a questionnaire-based approach known as contingent valuation (CV). By the end of the chapter, you will be able to identify environmental decision-making situations when the use of CBA/CV is justified, and situations when it is not. You will have been able to reach your own conclusion about an issue at the heart of debates about environmental valuation: are all environmental values reducible to a single monetary figure, or are these different kinds of values incommensurable – meaning that one kind of value cannot be expressed in terms of another?

Activity 7.1

Re-read the Pevensey discussion above. Can you see how incommensurable values are central to the group discussion?

Comment

Meg believes that the emotions people feel in knowing and loving their place cannot be expressed in money terms. Malcolm agrees with her, suggesting that human dependency on nature is of an entirely different order from anything associated with the exchange of money. Meanwhile, Carol believes a mercenary approach is characteristic of society today, and she strongly disagrees with it. For Meg and Malcolm, a sum of money could never substitute for personal attachments and the intrinsic importance of nature.

2   Environmental decision making

As you have learned from previous chapters in this book, environmental conflicts often arise from the actions taken by organizations and individuals. Some impacts, such as the building of a wind farm, will be immediate, tangible and local in scale. Other impacts, such as the liberalization of international trade, will be long term, intangible and global in scale.

Actions arise from decisions: choosing one strategy over another; weighing the costs and benefits of different combinations of activities; evaluating the claims of one interest group against others; and predicting the likely outcomes from different plans and projects.

In the Pevensey extract, this is what Christopher is driving at. There are experts ('they') who can look objectively at the scientific importance of one site and weigh its claims for importance against that of other sites. The rest of the group agree with him in the sense that they do believe that the experts need accurate information from the public – 'it comes down to answering this (questionnaire) honestly'. So the group are caught in a double bind. Most do not accept that it is right to give a money figure for what nature on the Levels is worth to them as individuals. But they have been asked to do so in a questionnaire that has official authority and that they feel must therefore be asking reasonable questions. We shall return to this issue later in the chapter.

We are all decision makers, and although our individual choices may seem paltry and insignificant in comparison to those of national governments or international agencies, there is much in common. All decision makers require knowledge and a basis for understanding the choices they need to make. Different kinds of knowledge and understanding are needed, especially when the decision has a substantial environmental component.

- *Scientific knowledge* is required to help define the problem, the range of technological interventions that may be feasible, the probability of success, and the possible risks that might arise at some point in the future.
- *Economic knowledge* is necessary to establish the range of costs and benefits associated with different courses of action and how they are apportioned between different interests, and the financial or commercial risks associated with different strategies.
- *Practical (commonsense) knowledge* is also needed. This is the knowledge and understanding that comes from everyday experience. As you will know from your own experience, decisions that are novel or unusual require much more thought and engender much stronger feelings of uncertainty and anxiety than do decisions that are familiar and routine.

At the same time, all decision makers are faced with constraints of different kinds. In this book, we have often described these constraints in terms of the relative power and influence that individuals and organizations have to take action in specific contexts. So, for example, in a contested decision about whether

Universal Studios should be allowed to build a theme park on Rainham Marshes (see Figure 7.3), another unloved patch of wetland just east of London's docklands, money talked in all sorts of ways: the developers boasted about the potential contributions the scheme would make to the UK's economy; the number of jobs it would create; the investment it would stimulate. The developers flew government ministers to Florida to inspect the Universal Studios theme park; paid for elaborate PR events; held press conferences and produced glossy brochures. Money meant the developers could rent offices close to Parliament: when environmentalists managed to get some critical media comment about the scheme, the developers could pop round to ministers and present their arguments without going public (see Harrison and Burgess, 1994).

Figure 7.3 Rainham Marshes – another unloved patch of wetland?

2.1 Ownership of and responsibility for land

Weighing up the pros and cons, or costs and benefits associated with different courses of action (or inaction), in whatever terms those costs and benefits are expressed, is fundamental to decision taking at any level. But, as we have seen throughout this book, environmental decisions are especially complex. This is because nature and the environment are a collective, common good as well as bundles of privately owned goods. So, in the case of Pevensey Levels the wetlands are farmed by approximately 60 farmers and tenants, some of whose families have worked their land for many generations. At the same time, the landscape and its nature matters enormously to many people who live in and around the Levels. To take another example, as you have learned in Chapter One, individual farmers were licensed to grow GM crops on their own land. Whether farmers grow oil seed rape or sugar beet or rear sheep or cows is a private, economic decision.

But the spread of GM pollen to other fields, the impacts of toxic pollens on invertebrates and the food chain have consequences for all and are, therefore, a matter of public or common concern.

This double 'ownership' creates a very difficult situation for environmental policy makers who are trying to represent 'the public interest' – the collective or common good that resides in nature and environment – when private owners want to do something that will bring them personal benefit but may harm the common good. For example, farmers in a river catchment will each use the amount of fertilizer they believe will maximize crop yield. One outcome of these individual decisions may well be that the river itself suffers eutrophication as the excess nitrates diffuse through the catchment. The farmer gains the cash benefit of higher crop yields; nature and society as a whole bear the costs of the ecological damage and, potentially, the costs of cleaning up the pollution. Equally, for many decades coal-fired power stations were able to release pollutants into the air without having to install expensive equipment to scrub clean their exhaust gases. These pollutants have been acknowledged as the main source of 'acid rain' which caused die back of forests in Scandinavia, Germany and other European countries (see Figure 7.4). The history of environmental conflict is littered with similar examples of private profit at environmental and social cost.

Figure 7.4 The consequences of acid rain in the Czech Republic.

Economists label this shifting of the costs of dealing with the waste products from the industry to the environment and society as a process of *externalizing costs*. What would have been a significant financial drag on the profitability of the industry, needing to be dealt with internally as one of the costs of production, is 'externalized' to the environment. Such an action is *efficient* economically: it achieves maximum profit at minimum cost. The actual costs – of not cleaning up –

are carried by the environment itself (sometimes vividly described as the 'sink' for industrial pollutants) and by the wider society and future generations. The environmental justice movement, discussed in Chapter Six, arose directly in response to industries seeking to maximize profits by externalizing the costs of properly treating waste products to the environment and to disadvantaged social groups. In such extreme cases, an ethical value of fairness and justice is beginning to over-ride an economic efficiency value. But it is very much more difficult to win the argument when economic, instrumental values are confronted by what seem to be weak, subjective, local and individual expressions of attachment. No contest?

2.2   What are environmental values?

Academics from many disciplines including philosophy, economics, geography, sociology and policy studies have been locked in sometimes heated argument about environmental values, and how they might be given due weight in policy decisions. John Foster is an environmental philosopher who is very interested in values. He writes: '"Value" ... is a word with all the complexity of life itself. What we value, and how we value it, depends both on our values and on the value of things in themselves – and there are important differences of meaning distinguishing all these usages' (Foster, 1997, p. 2). As you will have understood from Chapter Two, we can broadly distinguish between two kinds of values:

- *Non-instrumental values* are the values that individuals tend to attribute to nature or certain kinds of landscapes or to particular places that reflect personal experiences, shaped by social and cultural values of society. In some contexts, or in relation to certain categories of living beings or non-living things, value may be based within a moral sense of duty and/or a fundamental belief in its intrinsic worth. Non-instrumental values construct a socially and culturally accepted code of conduct for each individual; what that individual believes to be right and proper behaviour in relation to others, and to nature.
- *Instrumental values* reflect the utility of a good or a service to the individual for the satisfaction of a personal need or want. In these cases, a sum of money may well be a satisfactory means of expressing use value. So, when you go to the supermarket to buy a loaf of bread, you know how much money you would be prepared to pay for the commodity and what constitutes 'good value' for you. Similarly, if you enjoy walking in the forest as a way to relax and forget the troubles of the week, your expenditure, in getting to the forest, reflects its instrumental value. Instrumental values privilege the rights of the individual to seek maximum happiness for themselves without necessarily having regard for the needs of others.

I find it helpful to think of values as the reasons given for actions. My values explain why I am willing to walk round my local park in the mornings picking up the beer cans that last night's revellers threw into the hedge. I care about the park, and it is important to me that the thoughtlessness of some does not spoil it for all. At the same time, my study window looks over the park and I can go and walk in it whenever I want. It's the major reason why I bought my house. So, both instrumental and non-instrumental values underpin my actions in picking up other people's discarded beer cans.

Activity 7.2

Think about your everyday activities such as shopping, going to work, or going out at the weekends. Choose two activities. Then try to decide whether the reasons you do that activity are mainly instrumental reasons, non-instrumental reasons or a mixture of the two. For example, do you buy Fairtrade coffee or tea? Is that because it's on special offer at the supermarket (which would be an instrumental value) or because you want to support small independent growers in developing countries (which would be a non-instrumental value)?

Summary

- Environmental decisions are based on scientific, economic and common-sense forms of knowledge.
- Power and influence are particularly important factors in contested land-use decisions.
- Environmental decisions are very complex because both private interests and public interests have to be protected.
- In a free or lightly regulated context, industries are able to externalize some of their costs by polluting the environment.
- It is difficult to make an argument against economically efficient uses of the environment on the basis of non-instrumental values.

3   Economic approaches to environmental valuation

> Until the economic value of environmental quality is an everyday feature of the way we compute progress and, more importantly, of the way we make economic decisions ... the environment will not be given a fair chance.
>
> (Pearce, 1993, p.3)

Much of the impetus behind the concept of sustainable development comes from anxiety about the inadequate protection given to environmental systems and resources by the market. New environmental policies and new kinds of regulations have been developed in an attempt to ensure that the environmental costs of economic activity are fully accounted for. To achieve this kind of accounting, it has been necessary to break down the many different ways in which the elements of environmental systems 'help' the progress of economic activity and human welfare. This way of thinking is expressed in terms of the 'goods and services' that the environment supplies. Nature and the environment is a complex system of interlocking commodities. So, for example, the biological and chemical processes in a river that deal with the outputs from sewage treatment plants are performing a 'service' to society that should, in principle, be costed on a balance sheet. To take a different example, the park at the end of my garden gives me personal satisfaction; as a 'positional good', the park also adds to the economic value of my house when I come to sell it.

A 'positional good' refers to the financial benefit that accrues from environmental or other characteristics of a location.

A specialist branch of economics devoted to the pricing of environmental 'goods and services' has developed since the late 1960s with major centres of research in the USA, Europe and Australia. The central challenge facing environmental economists is the fact that the great majority of environmental commodities, such as the capacity of water to absorb pollutants from industrial and agricultural activities, are not traded in a market. According to neo-liberal economic theory, as you saw in Chapter Five, each individual expresses a preference for a particular good by how much he or she is prepared to pay for that good. Where there is scarcity – a limited supply of a product and/or great competition between individual consumers – the price of a product will rise. The problem with the vast majority of environmental 'goods and services' is that there is no market and they are not traded. So how is a monetary valuation to be obtained and how should it be used in policy and project analysis? We will focus first on the most common and widespread methodological approach used by environmental economists to capture individuals' environmental values, before turning to consider how these values are used in policy making.

3.1 The concept of total economic value

Environmental economists begin the process of environmental valuation by distinguishing between two kinds of values, using a slightly different terminology from the one we have been using in the series. Environmental economic values are either:

- **Use values** arise from actual use of a particular environmental 'good' or natural resource. Buying and selling land is a prime example of a market-traded environmental commodity where price embodies the value of the resource for seller and purchaser. (We have been referring to these as roughly equivalent to instrumental values.)

  Use values arise from actual use of a particular environmental 'good' or natural resource.

- **Non-use values** acknowledge the values that environmental 'commodities' have values for certain individuals but do not have any market in which they might be traded. Blackbirds and thrushes are not tradable commodities now (although they were in eighteenth-century England) but are valued highly by, for example, members of the Royal Society for the Protection of Birds. The landscapes of the National Parks of the USA are not tradable but they do embody certain kinds of values, including a powerful resonance for white American cultural identity. (We have been referring to these as roughly equivalent to non-instrumental values.)

  Non-use values acknowledge the values that environmental 'commodities' have for certain individuals, even though they do not have any market in which they might be traded.

Both kinds of value are attached to environments and natural resources, and, added together, produce what is called **total economic value** or **TEV**. Two leading environmental economists, Ian Bateman and Kerry Turner (1993), teased out the different kinds of use and non-use values expressed in environmental resources. Figure 7.5 shows the TEV for woodland.

Figure 7.5 illustrates the different kinds of values that may be attached to woodland resources. The total benefits that a woodland contributes to human welfare is the sum of the development benefits (use values), and conservation benefits (non-use values).

Use values are classified as either:

- *direct use values*: with which an individual derives actual economic benefit through, for example, buying or selling timber; or as
- *indirect use values*: through, for example, paying for a holiday in a forest nature park where experiencing the forest is a significant motivation for the expenditure.

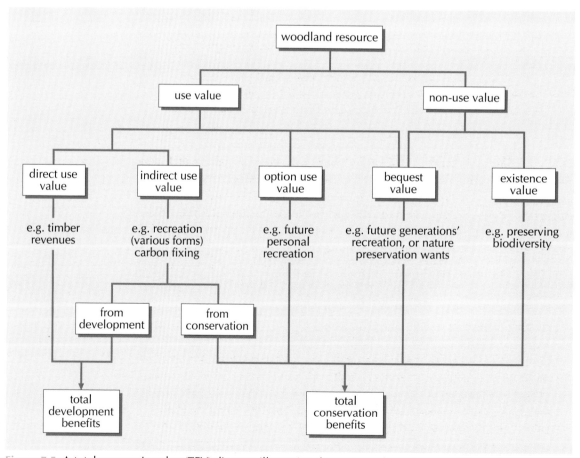

Figure 7.5 A total economic value (TEV) diagram illustrating the various elements comprising value.

Both streams of monetary use values add up to the total development benefits that accrue from the woodland resource. In Figure 7.5, two further kinds of use values signify individuals' future economic intentions in relation to the resource.

- *option use value*: which refers to the potential to capitalize on the resource at some future date;
- *bequest value*: which may be classified as either a use or non-use value depending on the context. As a use value, it represents the intention to hand the resource down to future generations for economic exploitation.

Two kinds of non-use values contribute to TEV:

- *bequest value*: represents the rights of future generations to experience the qualities of nature experienced now without converting them to cash, for example, by selling the timber;
- *existence value*: this is a way of expressing the intrinsic worth of trees in woodland ecosystems, and their right to exist and flourish independently of human beings.

Non-use values are tricky things for environmental economists to deal with. As Bateman and Turner write:

> Non-use values are more problematic. They suggest non-instrumental values, which are in the real nature of the thing but un-associated with actual use, or even the option to use the thing. Instead such values are taken to be entities that reflect people's preferences, but include concern for, sympathy with, and respect for, the rights or welfare of non-human beings.
>
> Bateman and Turner, 1993, p.122

As **Hinchliffe and Belshaw (2003)** point out, non-use values and non-instrumental values reflect ethical and/or aesthetic beliefs that provide the moral foundation for individuals' actions. For someone who holds all life sacred, to suggest that a sum of money can compensate for the destruction of a woodland is offensive in the extreme. Fundamental values about what is right or wrong in this moral sense are extremely difficult, if not impossible, to accommodate in an economic framework. And yet, they must somehow be taken into account if decisions are to be based on a full recognition of the personal, social and cultural importance of nature and environment.

Activity 7.3

Take an environmental resource and see if you can draw out the different use and non-use values in a similar way to that done in Figure 7.5. In particular, see how many different kinds of use and non-use values you can identify for the resource. For example, think about a resource that has a clearly identified market value (such as powerful waterfalls) and another resource for which society's values have clearly shifted (think about the TEV for whales in the nineteenth century, and what their TEV would be now, Figure 7.6).

Figure 7.6 The TEV of whales has changed as time has gone by (a) a whaling ship in the nineteenth century and (b) a modern tourist boat with fee-paying whale-watchers on board.

3.2 Measuring costs and benefits in environmental policy appraisal

Cost–benefit analysis (CBA) This is the major technique used to help appraise how efficient and how effective different strategies might be.

Computing TEV is a necessary part of doing a **cost–benefit analysis (CBA)**, which is the major technique used to help appraise how efficient and how effective different policies might be. Over the last 30 years, CBA has become the standard methodology used by experts in environmental decision making (Pearce, 1998). For example, whenever the government has to take a major investment decision that is likely to be contested by different interests, it will require civil servants (supported by experts from many different disciplines) to complete a policy appraisal. One well known case is the question of whether and where to allow expansion of airports in south-east England: should there be a third London airport built on Maplin Sands or the North Kent marshes? Should Gatwick be expanded further? The inquiry to consider building another terminal at London Heathrow began in 1995 and took six years to reach its decision to approve Terminal Five.

Table 7.1 shows how the then Department of the Environment (DoE, now DEFRA) uses CBA as an integral part of its policy appraisal process (see Figure 7.7).

The critical issue in terms of completing a CBA is to ensure that as many of the direct and indirect costs and benefits (including matters such as loss of wildlife or landscape amenity) are expressed as sums of money that can be included in statistical analyses of the performance of different options.

In a paper defending the use of CBA in environmental decision making, Hanley (2001) argues that CBA has many interesting and important things to say about a project. For example:

- CBA tests for *economic efficiency* in resource allocation. CBA provides an explicit rationale for making a choice between different actions. In deciding to allocate money for one project, other opportunities for spending that money

Table 7.1   How the Department of the Environment used CBA as a part of their policy appraisal

Steps in policy appraisal

- Summarize the policy issue: seek expert advice to augment your own knowledge as necessary.
- List the objectives: give them priorities; identify any conflicts and trade-offs between them.
- Identify the constraints: indicate how binding these are and whether they might be expected to change over the period of the negotiation.
- Specify the options: seek a wide range of options (including do nothing) and continue to look at new options as policy develops.
- Identify the costs and benefits, including the environmental impacts; do not disregard likely costs or benefits simply because they are not easily quantified.
- Weigh up the costs and benefits, concentrating on those impacts that are material to decisions.
- Test the sensitivity of the options to possible changes in conditions, or to the use of different assumptions.
- Suggest the preferred option, if any, identifying the main factors affecting the choice.
- Set up any monitoring necessary so that effects of policy can be observed, and identify any further analysis needed.
- Evaluate the policy at a later stage and use the evaluation to inform future decision-making.

Source: Department of the Environment, 1991

Figure 7.7 The cover of the Department of the Environment's policy appraisal book.

are closed off. As you will know from your own experience, financial resources are finite –the great night out will probably mean baked beans for the next fortnight! This is the **opportunity cost** you have chosen to bear. CBA is concerned with two issues in particular: to ensure that the money spent on any project is the lowest possible for the level of benefit secured; and to be sure that the opportunity costs are not excessive in relation to the benefits of the policy. Furthermore, CBA forces decision makers to think systematically about all the potential costs and benefits of a policy and to do so in such a way that new information is made available to them.

Any economic choice involves consideration of alternatives. When a choice is made the other alternatives are lost or foregone. In economic terms this is regarded as the opportunity cost.

- CBA allows the *integration of financial and environmental effects* in decision making, which ensures that the consequences for the environment of any decision are fully identified and acknowledged in the trade-offs between different choices. This, if you refer back to Pearce's quote at the head of this section, is the specific goal of environmental valuation. To protect the environment is vital and only by allocating monetary values to the environmental impacts of specific actions can the environment be brought fully into decision making.
- CBA *incorporates social values* by bringing in values other than those of experts into the decision framework. Until relatively recently, CBA has been the exclusive concern of experts (economists, civil servants and project managers, among others). Now, through the use of social surveys, the preferences of individual members of the public can be measured and incorporated in the CBA. This is particularly important because it allows

decision makers to claim that CBA is a democratic process in which the views of members of the public are taken into account.

So, let us move forward to explore the major technique used to measure public values for environmental 'goods and services' in an effort to create democratic, representative CBAs.

3.3 Contingent valuation: an overview

Contingent valuation   (CV) methodology has become the major survey technique through which environmental economists attempt to gain a representative sample of monetary values for specific environmental 'goods and services'. Unlike tradable commodities, it is not possible to measure actual consumer behaviours in a marketplace. CV methodology sets a hypothetical market, in order to measure individuals' stated preferences for environmental 'goods and services'. In other words, individuals are asked what they would do, *if* there were an opportunity to trade in the particular environmental 'commodity'. So, for example, how much would you be willing to pay if there were a market for cleaner air in your neighbourhood? What would you personally be prepared to forego in order to 'buy' the cleaner air?

Let us pose this in a more tangible way. There is a major housing shortage in south-east England. Lowland heaths are also scarce, and much has been lost to development over the last 50 years (see Figure 7.8). A private house-builder has applied for planning permission to build on heathland adjacent to the built-up area of Bournemouth. The local authority conducts a CV survey to establish the value of the lowland heath to members of the public.

Just step back a little. As a consumer, you engage in markets in one of two ways, either as a buyer or a seller: you spend money to obtain the benefit of something you want or you receive money for something that someone else wants from you. CV needs to make you into a 'buyer' or 'seller' of the values in the heathland. How can it do this? In one of two ways: it can ask you how much you would be willing to pay for the heathland to be saved from development (i.e. it puts you in the position of a buyer in the market for heathland); or it can ask you how much compensation you would need before you would be willing to have the heathland destroyed (i.e. you 'sell' your value in the heathland to the developer).

CV describes these two trading options in the following way:

- *Willingness to pay (WTP).* This figure is the amount of money that an individual would hand over to obtain the marginal welfare gain of an environmental 'commodity'. So, for example, in response to a CV survey I might say that I would be willing to pay £25 a year to a charitable trust fund to support the re-instatement of wildflower meadows in south-east England. Economic theory assumes that I have a mental account of my annual income and the demands on that income. In expressing my WTP, I have made a

Figure 7.8 Lowland heath housing in Dorset.

rational decision about what to forego in 'spending' money on this 'environmental commodity'.

- *Willingness to accept (WTA)*. This figure is the amount of money that an individual would accept in compensation for the reduction or loss of an environmental 'commodity'. Neo-liberal theory assumes that I can be compensated for the loss of wildflower meadows through receiving a sum of money that I could then, hypothetically, spend on satisfying other wants and needs. In other words, I can substitute the loss of the wildflowers by buying something else that will give me the same level of satisfaction. There is evidence to show that WTA figures are much less reliable than WTP. Individuals find it more difficult to produce a meaningful sum of money in these contexts. It may also be likely that the task poses an immediate ethical choice that individuals refuse. For these reasons, most CVs use the WTP option.

That is the theory, so how does it work in practice? The CV questionnaire typically has three components:

- *The scenario*: is a written description (sometimes augmented with visual material) that describes the 'commodity' and the hypothetical market the individual is invited to trade in.
- *The payment questions*: ask, in different formats, for a WTP/WTA figure.
- *Respondent information questions*: in addition to basic socio-economic data, these may include questions about environment attitudes, current use of the 'commodity' and patterns of household expenditure. This information allows for analysis of possible relationships between individuals' willingness or refusal to participate in the contingent market and the WTP figure.

Representative samples of respondents are drawn from the population and the CV survey is completed by asking each respondent to answer the respondent information questions, to read the scenario and then respond to the WTP question. The WTP question may be phrased in terms of discrete sums of money ('would you be willing to pay £5; £10; £15 a year' and so on) or as a continuous amount ('how much would you be willing to pay a year?'). In 'best practice' CV surveys, the questionnaire is completed through face-to-face interviews. Analysis includes calculating the average WTP figure. This means adding all the individual bids together and dividing by the total number of respondents to get a mean WTP. Extreme bids are excluded to avoid distorting the average estimate: think of the impact of one person saying they would pay £1 million when everyone else offered around £10. The averaged WTP figure is then included in the CBA. Well over 2,000 CV studies have been carried out in the last 20 years or so and a body of good practice continues to develop.

Activity 7.4

It is challenging to get to grips with a quite complicated set of arguments and a novel methodology. Having worked through Section 3, it might be helpful to complete an actual CV questionnaire for yourself. You should answer the questionnaire in Figure 7.9 (on the opposite page) by yourself first. Then if you wish, try it out on friends and relatives.

This questionnaire was designed to be completed by students, by Dr Henrik Svedsater, a social psychologist at the London School of Economics, as part of his PhD research into understanding CV decision-making processes.

Before you read on, make sure you have answered the questions in Figure 7.9. This is a standard CV protocol that seeks to measure the non-use value that you would receive from contributing to the WWF fund for saving African elephants. 'Non-use' in this sense means that you would not gain instrumental value from seeing the elephants in the wild. Rather you have used a sum of money to express the benefit that you would receive by knowing that an (unspecified) number of elephants might survive whereas they might not without the WWF project. If we were to work out the average WTP figure for all the students taking the course, and incorporate that into an analysis of all the costs and benefits of the WWF project, it would be possible to determine whether the project was economically efficient or not.

Summary

- To achieve greater environmental protection, it has been necessary to think about environment systems as providing 'goods and services' for human use.
- Total economic value is the sum of all the use and non-use values provided by environmental 'commodities'.
- CBA is the standard decision-making tool used by experts in environmental policy appraisal.
- CBA has three important characteristics: it supports economic efficiency; it integrates financial effects with environment effects; and it incorporates social values into environmental policy appraisal.

Saving the African elephant

Please answer all questions.

1   What is your sex?
   ☐ Male
   ☐ Female

2   Which age group do you belong to?
   ☐ 18–25
   ☐ 26–34
   ☐ 35–49
   ☐ 50–64
   ☐ 65 +

3   What is your monthly gross income, before any expenditure?
   ☐ £0–£499
   ☐ £500–£749
   ☐ £750–£999
   ☐ £1000–£1499
   ☐ = +£1500

4   Do you have any children living at home with you? If so, how many and what are their ages?

5   Are you a member of any of the following environmental organizations?
   ☐ Worldwide fund for nature
   ☐ Friends of the Earth
   ☐ Greenpeace
   ☐ RSPB
   ☐ Other (please specify)

This study is focused on saving the African elephant which is an endangered wild animal.
Please read the text below carefully before answering the following questions.

Measures to save the African elephant

The long-term survival of the African elephant is cause of great concern. The number of elephants has fallen dramatically during the second half of the century. In 1979, there was an estimated 1.3 million elephants in Africa, but by 1995, this figure had shrunk to around 400,000. Part of the decline is due to the availability of new dry-land crop strains, with the consequence that former elephant range-lands are now being cultivated. Furthermore, in forest areas the impact of major logging programmes is opening up and destroying elephant habitat. Apart from such widespread changes in the extent and pattern of land use, a major cause of the decline is poaching to satisfy demand for ivory and recreational, illegal hunting.

6   What did you previously know about this problem?

   ☐ I knew the African elephant was a threatened animal
   ☐ I have heard about the problem but did not know much about it
   ☐ I did not know that the African elephant was a threatened animal

Figure 7.9 A contingency valuation questionnaire on the African elephants.

As a consequence, approaches are needed to stop the decline in the number of elephants. Apart from traditional anti-poaching efforts and the elimination of market demand for ivory products, it is essential to ensure the survival of the remaining herds. The WorldWide Fund (WWF) is a major actor in this field. It is currently running a campaign by setting up and managing reserves in order to protect wild elephants. Experience has shown that local involvement is important in these attempts, such as community based management, whereby landowners share responsibility for and benefits accrued from elephants.

However successful these conservation approaches may be, they bear significant costs and as the economic situation in many third world countries continues to decline, wildlife departments and local communities are suffering significant budget cuts. This makes international support for elephant conservation more important than ever. In this study we are interested in knowing how much the efforts to save the elephants are worth to you. More specifically, we would like to know how much you are willing to pay, as a yearly contribution, to support the WWF campaign.

In the question below we want you to state how much you are willing to pay as a contribution to the WWF campaign to save the elephant. The willingness to pay is an annual payment. Take your time and try to consider the following before answering:

- Your income
- Your current expenses
- Your possible future use of your income

7   My maximum willingness to pay for saving the African Elephant is £ .... yearly

Follow-up questions

8   Please state the most important reason for the amount that you have agreed to pay. (Tick one option only)

☐ It is what saving the African elephant is worth to me

☐ I cannot afford to pay more

☐ Based on the average contribution, I think this will be sufficient to cover the costs

☐ I believe this is a fair amount given my own responsibility for the problem

☐ I believe this is a reasonable amount considering what other people would pay

Other consideration(s)

9   To what extent do you agree or disagree with the following statements:

'People have to make choices between environmental issues and economic development'

☐ strongly agree   ☐ agree   ☐ undecided   ☐ disagree   ☐ strongly disagree

'To make decisions and priorities about this and other similar environmental problems on the basis of citizens' willingness to pay is an appropriate and sensible approach'

☐ strongly agree   ☐ agree   ☐ undecided   ☐ disagree   ☐ strongly disagree

End of questionnaire

- Contingent valuation is a social science methodology designed to elicit instrumental and non-instrumental values for environmental 'commodities'. It is based on a hypothetical market for an environmental 'good or service' in which individuals are invited to trade.
- Trading in a CV market consists of saying how much one is willing to pay to save a threatened environmental 'commodity', or how much one is willing to accept in compensation for its loss. WTP is used more often than WTA because it appears to be more reliable.
- CV is able to produce a statistically representative average WTP/WTA figure, which is used as an aggregate measure of the instrumental and/or non-instrumental values in a CBA.

4 What does the WTP figure mean to respondents?

CV surveys can be about matters that are closer to or further away from individuals' everyday experiences. If you are an ardent backpacker who walks the Lake District hills every weekend, you will have substantial knowledge of just how much it costs to get to the Lakes, how much to stay in a hostel, how much to pay for new boots. So, if I ask you, as part of a CV survey, how much you would be willing to pay to set up a Trust Fund to protect the footpaths of the high hills from further erosion, you would be able to give me a figure based on your use value, which would probably be a fairly accurate estimation of how much of your discretionary income you would be willing to commit.

But suppose you have no personal experience of the Lake District, no means of knowing the utility of the hills in terms of the costs and benefits of hill-walking. In responding to the same CV survey, what would your WTP figure mean? Would it be an economic value? – that is, what you would actually choose to pay if push came to shove? Or would it be a political gesture – that is, that you believe some organization should pay to protect the hills? Or would it be something other than that: perhaps a number that gave you a small warm glow of satisfaction that you have responded altruistically to a questionnaire survey – 'to impress the interviewer' as Meg said in the Pevensey discussion at the start of the chapter?

Pause for a minute: can you identify the motivations that led you to offer your WTP to save the African elephants? A number of studies over the last few years have tried to find out what is going on inside people's heads when they answer a CV survey. I want briefly to discuss three of them. Apart from general curiosity about the process, there is a specific problem. It seems that respondents may not behave 'economically' (that is, as rational, utility-maximizing actors in a market). If I know I can buy six cakes for £1, then I would expect to pay £2 for 12 cakes or 50 pence for three. But this rationality does not seem to work well in hypothetical, environmental 'markets'. Respondents do not seem willing to pay proportionately more for a greater amount of the 'good'.

4.1 The Central Flyway CV survey

One excellent example of this problem is the experimental CV survey (Boyle et al., 1994) in which respondents were asked to give a WTP to reduce migratory bird deaths.

The hypothetical scenario concerns the preservation of migratory waterfowl using the Central Flyway of the USA (see Figure 7.10) by preventing them from dying in settling ponds that contain waste-oil. There are about 250,000 of these ponds in the remote and sparsely populated areas of Texas, Oklahoma and New Mexico. The ponds vary in size from around 3 metres to 30 metres in diameter and contain water, oil and other by-products from oil and gas drilling operations (see Figure 7.11). These by-products often kill the waterfowl that land on the ponds. The proposed solution was to enact government regulations requiring oil and gas producers to cover the ponds with fine mesh wire that would prevent the birds from coming into contact with the oil. Compliance would be enforced by the Fish

Birds fly south via the Central Flyway to overwinter in the Gulf of Mexico. They then fly back up to the Arctic at the beginning of the breeding season.

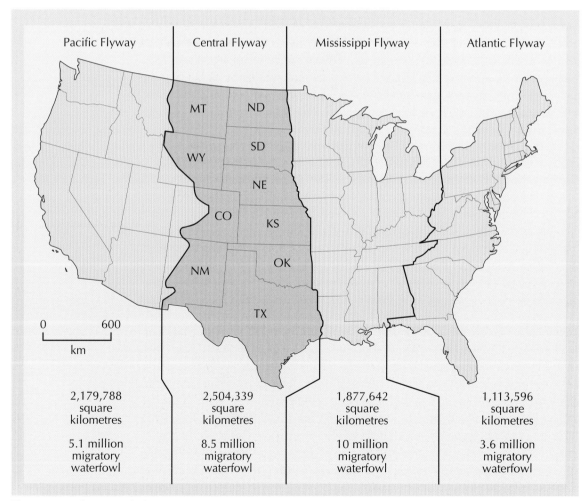

Figure 7.10 The Central Flyway, one of the many bird migration routes down through the North American continent.

Figure 7.11 An example of a settling pond on the Central Flyway. These ponds often contain waste oil from drilling operations. Migrating birds will fly over these ponds and try to settle on the surface. Some oil companies are responding to this situation and this image shows a site that is now being managed for nature conservation

and Wildlife Service regulations of the USA. The cost of the solution would be passed on to consumers in the form of higher prices for petroleum-based products.

Three versions of the scenario were produced, each with a different number of birds saved: 2,000, 20,000, and 200,000. Both the absolute number and its percentage of the total number of migrating waterfowl were included for each level: 2,000 = 'much less than 1 per cent' of the total number of birds using the flyway; 20,000 = 'less than 1 per cent'; 200,000 = 'about 2 per cent' of the total. So, one scenario read as follows:

> In 1989, for example, about 20,000 migratory waterfowl died in these holding ponds. This was less than one per cent of the 8.5 million migratory waterfowl in the Central Flyway. The affected migratory waterfowl include mallard ducks, pintail ducks, white-fronted geese, snow geese, and greater sandhill cranes.

To eliminate respondents with actual use-value for the birds (visitors and hunters, for example), the survey was carried out in shopping malls in Atlanta, Georgia. In total, 1,205 people completed the CV survey, divided into three sub-samples for the three different scenarios. The average WTPs under the three scenarios were $80, $78 and $88 for preventing 2,000, 20,000 and 200,000 bird deaths respectively.

sandhill crane (*Grus canadensis*)   white-fronted goose (*Anser albifrons*)

snow goose (*Anser caerulescens*)

The middle figure ($78) highlights the problem because it tells us that the average WTP for saving 20,000 birds was less than that for saving 2,000 birds. In other words, the respondents were not behaving in an economically rational way, because they seemed to be willing to pay less for saving more birds. Or, to put this another way, they were willing to pay more to save fewer birds.

This result, which has been replicated in other CV surveys, has prompted researchers to ask what is going on inside people's heads when they produce a WTP figure. Schkade and Payne (1994), using the Central Flyway scenario, asked 120 volunteers in a laboratory-based experiment to verbalize their thoughts as they worked through the CV. The WTP figures obtained were very similar to those from the original study, suggesting that the protocol itself was robust. Very few statements about the economic or monetary value of the birds were made as people verbalized what they were thinking. There were very few examples of a trade-off between the money forgone and the birds saved. This contrasts with CV studies of familiar market commodities, which show that individuals do think economically about those 'goods'. The transcripts obtained by Schkade and Payne show individuals drawing on economic, political and social reasoning. For example, 41 per cent acknowledged that something should be done about the problem and then tried to decide how much money would be needed *if* every household contributed. Here is an example of the reasoning of one individual in the study:

> Um, this is very difficult to determine. You have to consider how many millions of people in the country would also be contributing to this ... as far as how much per family this would break down to and what is the cost of putting this netting over all the different ponds, and how many ponds there are, so that would be the cost. I mean, if it ... comes out to be a couple of dollars per household, then it seems reasonable. If it comes out to be something more than that, um, it seems a little high.

Approximately 23 per cent accepted that it was inevitable consumers would have to pay, and then worked out an appropriate amount. Twenty per cent of the sample said they just guessed or made a number up. A further 7 per cent saw their WTP as a contribution to charity and used that as a basis for a figure, while 23 per cent were concerned about broader environmental issues. Their WTP was a political gesture in an ethical debate. For example:

> I'm thinking like, um, I would pay $50, maybe even $100, $50–$100 a year would sound about right to me. I think it's a little bit high, but I kind of think it's important because, um, I feel it's important for us to preserve the wildlife, and not, not only ducks and geese but other animals too. I feel like if we just continue to let things go, we're going to be paying money for other things that may not be quite as important, and when you're killing off all of these animals, and I don't think that's good. I'd like my children to see these animals one day ... when I have them.

In these two extracts, the first person is reasoning through instrumental values to find a WTP figure that would seem reasonable to her. In the second, we have an example of reasoning through non-instrumental values. In both cases, the respondents have attempted to find a monetary sum that can represent their value. Neither have challenged the basis of the CV exercise. But let us now, finally, turn to a case study where members of the public were given the freedom to question the basis of the CV survey as well as to explore what their WTP meant to them.

4.2 The Pevensey Levels CV survey

The two American studies just discussed were purely academic. But CV surveys are also used quite widely as part of CBAs in real policy contexts. In these cases, the claim that CV provides a way of exercising 'economic democracy' becomes significant. Just consider the range of expertise needed to capture environmental values in the CV/CBA methodology. Academic economists and experimental and social psychologists determine the technical issues; civil servants and business managers understand the extent to which they can use the results; trained economists within organizations carry out the studies; and these experts, together with their board members and/or political masters decide whether it was all worth it. The appraisal loop is neat and largely impenetrable to anyone without the necessary power and expertise.

pintail duck (*Anas acuta*)

Experts may claim democratic legitimacy, saying that the values of the general public have been captured in 'state of the art' CV surveys. The extent to which this confidence is misplaced can best be revealed through qualitative research, preferably with respondents to CV surveys. Do members of the general public understand the task being set them in a CV survey? Are they content that their values can be expressed in a monetary figure? To what extent do they know how their answers will be used and are they willing to have the answers so used? The Pevensey Levels study that my colleagues and I carried out from 1993 to 1995 attempts to answer these questions (Clark et al., 2000).

We wanted to discover what respondents to a CV survey thought about the validity and legitimacy of the approach for capturing their values for nature and to explore the extent to which members of the public (rather than experts) were persuaded that CV/CBA was 'the only show in town' (i.e. were there alternative ways to make decisions about environmental resources, ways that provided space for ordinary 'lay' people to contribute meaningfully to the process?). English Nature, the government organization with responsibility for nature conservation in England, agreed to work with us on the research by commissioning environmental economists from Newcastle University to conduct a CV survey of their 'Wildlife Enhancement Scheme' (WES) policy on the Pevensey Levels. The WES was introduced by English Nature in 1992. Farmers and landowners entering land in the scheme received a payment of £72 per ha. per year to maintain their existing grazing practices, and to follow a management plan as prescribed by English Nature to maintain the wetland habitat for its conservation interests (Willis et al., 1996).

The economists designed a CV survey and Figure 7.12 (on the opposite page) shows you the scenario that was used in the survey. In order to measure use values, face-to-face interviews were conducted with both households in or on the boundary of the Levels and with those visiting the Levels. In order to measure non-use values, interviews were conducted with households living up to 60 kms away who had not visited the Levels in the previous year. We then recruited members for two in-depth discussion groups from among respondents to the CV survey. One brought together people living in and around the edges of the Levels (the Pevensey group); the other people living approximately 50 km east of the Levels in Maidstone, Kent (the Maidstone group). Recruitment was based on the range of WTP figures. The groups followed the same agenda over their four weeks, as summarized in Table 7.2.

Table 7.2   The sequence of group discussions

Schedule	Agenda for the in-depth discussion groups in Pevensey and Maidstone
Week 1	Introductions, the locality, nature and place-based interests. Members given scenario to take home for the week.
Week 2	The information scenario: each member of the group invited to describe her/his own reactions to the scenario, before opening up for general discussion. Members given copy of CV questionnaire to take home for the week.
Week 3	The questionnaire: each member of the group invited to describe her/his own reactions to the questionnaire, before opening up for general discussion.
Week 4	The WES, local environmental politics, and winding up.

Notice that the group members each had an opportunity to consider the information scenario and the questionnaire at their leisure, before the group discussions.

Tables 7.3, 7.4 and 7.5 summarize members' comments about their WTP figure and the process of monetary valuation. (Table 7.5 is taken from a discussion group of visitors to the Levels.)

Activity 7.5

Read the scenario (Figure 7.12) and then read through the comments in Tables 7.3, 7.4 and 7.5 carefully. Answer the following questions before reading on for a discussion of the groups' comments:

- What evidence is there that individuals have used instrumental values as the basis for their WTP?
- Have they engaged in some kind of mental accounting to produce a monetary value – even if they may not be comfortable with it?
- What evidence is there that non-instrumental values were important for the participants? Is it possible to describe a common or shared ethical position?

An Introduction to The Project

- The Pevensey Levels have been designated as a Site of Special Scientific Interest (SSSI) by English Nature, the official Government advisory body for nature conservation in England.

- This map shows the area of the Pevensey Levels. The Levels cover nearly 9,000 acres (about 14 square miles) of wet grassland and marshland between Eastbourne and Bexhill.

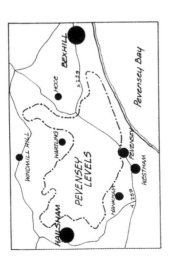

- English Nature is running a pilot Wildlife Enhancement Scheme on the Levels, and is paying farmers and landowners to manage the Levels to help conserve and enhance the wildlife.

- As part of a review of the pilot project, English Nature would like your views on how much you value wildlife conservation generally and on the Levels in particular.

- Your views will be treated as confidential and no personal details will be revealed to anyone else, unless you agree to take part in a second stage of the project which the interviewer will describe. Unless you give your consent, neither your name nor your address will be added to the questionnaire.

Wetland Habitats in South East England

The Pevensey Levels is one of more than 3,700 Sites of Special Scientific Interest throughout England, covering semi-natural habitats which include:

- coasts, cliffs and beaches
- woodlands
- meadows, fields and heaths
- rivers, lakes and ponds
- wetlands (fens, bogs, marshes)

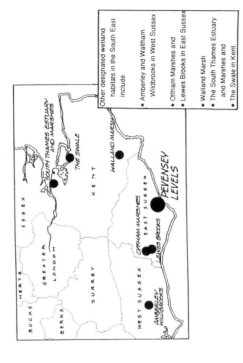

During the last 50 years nearly three quarters of lowland wet grasslands and similar habitats have disappeared in the South East, due to agricultural intensification, development and other changes.

Figure 7.12 The four-page booklet that presented the scenario for the Pevensey Levels WES CV survey.

1

2

The Future *With* the Wildlife Enhancement Scheme?

The main objectives of the scheme are to encourage farmers and landowners to farm the land in a traditional and sensitive way which safeguards the wildlife. This photograph shows a well-maintained ditch which is rich in wildlife.

What should the Wildlife Enhancement Scheme Achieve?

The result over the longer term will be to conserve the variety of wildlife and could lead to its enhancement.

Wildlife Interests on The Levels

Traditional management practices have allowed the wildlife variety to flourish: this is just a summary of the many rare or uncommon species found on The Levels:

Dragonflies
eg Hairy Dragonfly

Fen raft spider

Flowering plants
eg Arrow Head
Water Soldier
Marshmallow

Water Snails
eg Ramshorn Snail

Water Beetles
eg Great Silver
Water Beetle

Song Birds
eg Yellow Wagtail
Sedge Warbler
Reed Warbler

Wading Birds
eg Lapwing
Golden Plover
Snipe

Birds of Prey
eg Hen Harrier
Peregrine Falcon
Short-eared Owl

3

The Future *Without* the Wildlife Enhancement Scheme?

Agricultural drainage, re-seeding, reclamation, lack of ditch management and intensification all contribute to a slow but steady process which reduces the wildlife of the Levels. This photograph shows an unmanaged ditch which is poor in wildlife.

What will happen without the Wildlife Enhancement Scheme?

If recent trends continued, without management for conservation, the wildlife variety of the Levels would change and some species could become much less common or disappear altogether.

Wildlife Interests on The Levels

Dragonflies   ?

Fen raft spider   ?

Flowering plants   ?

Water Snails   ?

Water Beetles   ?

Song Birds   ?

Wading Birds   ?

Birds of Prey   ?

4

Figure 7.12 continued

Table 7.3   Summary of residents' comments on the Pevensey CV survey (from the Pevensey residents discussions)

Group member	WTP[1]	Comments made in the group discussions
Christopher (age: 25–34; civil engineer; £20–30k annual household income)	refused	I struggled with the money business because there are so many competing claims and I've come to the view that I need to contribute nationally and then have somebody to even it all out for me. I didn't like the questions. I wasn't happy answering them. I couldn't place what it was really asking. Someone has to make decisions but it shouldn't be on monetary value alone. It's to make sure the wildlife goes on surviving, regenerating, giving green space for the earth to develop, rather than just for us to go and enjoy it.
Carol (age: 45–54; parish clerk; £7.5–9k annual household income)	0 x	I said I couldn't afford to pay anything at the moment because of our financial situation but that's not a measure of how important the Levels are to me. We cannot put a price on nature. It's priceless.
Barry (age: 45–54; retired automotive engineer £15–20k annual household income)	2 x	I didn't think much about the money question, because to me that was an open your cheque book and sign a blank cheque. It's a totally disgusting idea. You can't put a price on the environment. You can't put a price on what you are going to leave for your children's children.
Kathy (age: 35–44; youth employment trainer; £30–40k annual household income)	3 x	That few pence a year and how many times – I found that really difficult. But I think it was what I could afford. If you pay money you have the right to see where that money is being spent, and is it being spent wisely. We have to make choices but on the basis of full knowledge. You can't judge one case in isolation.
Greg (age: 45–54; electronics engineer; £10–15k annual household income)	4 x	It was such a small amount. I mean, two times what? Nature is not ours to sell. The only answer is to become a much more political animal and bend the ears of MPs so the government would know that there would be such an uproar.
Meg (age: 45–54; social work administration; £10–15k annual household income)	10 x	I said an amount for me but I would have probably said that I wouldn't mind paying fifty pounds a year but to do it by how many times I found really difficult. It fills me with horror to think we all might have been unemployed and said we couldn't have afforded to pay even if we wanted to. Putting a price on nature is immoral. Species are irreplaceable.
Malcolm (age: 45–54; carpenter; £15–20k annual household income)	100 x	I thought you've got to think in terms of everyone's five pounds. If everybody in the area paid five pounds the farmers are going to get a lot of money. That way of thinking, it's not really on is it? You can't put nature on the stock market. It's our very existence. It's our future. Any open space should be valued. It should be legislated, so decisions are made that way.
Keith (age: 55–64; self employed agricultural engineer; refused to state household income)	1000 x	I didn't really understand what she was on about to be honest. I don't mind paying my fair share but I'd much prefer to pay somebody who's going to do something with it. I think this taxing from individuals won't work. We would do better to get the money for the conservation people to purchase the marshes. You can't trust the government to protect nature, even when you do pay taxes for them to do so.

1. The cost of implementing the WES came from English Nature?s block grant from Government and, therefore, from income tax. Averaged over all tax payers, it amounted to ?a few pence per household per year?. Respondents to the CV survey were asked to say how much more they would be willing to pay e.g. twice as much, three times as much, and so on.

Table 7.4   Summary of individual non-residents' comments about their WTP figures and monetary valuation, extracted from the Maidstone residents (MRG) discussions

Group member	WTP figure	Comments made in the group discussions
Ray (age: 55–64; retired civil servant Ministry of Defence, £10–15k)	refused	I thought that's a question that cannot be answered. I didn't answer it. I needed to know a lot more. To ask *your* specific interest, and what *you* would pay, I don't think most people often think about nature in those terms. Nature's part of the web of human life; so it's very difficult to put that sort of monetary value on it.
Laura (age: 45–54; unemployed; £20–30k)	0 x	I was surprised to find that money was going to this scheme. My first inclination was to say no. I think it needed a more direct approach, don't hoodwink us with this kind of questionnaire. It's the value of the countryside but the actual cost of it can't be defined.
Carla (age: 35–44; primary school teacher; £20–30k)	2 x	I found the money question impossible to answer. I tried to be positive, I think I did try to gild the lily a little. You don't know what's on the next page: would you put your hand in your pocket for this, and this, and this? You can't take one area in isolation very easily. We have to prevent the loss of species and stop being selfish.
Bob (age: 45–54; retired sales representative, £15–20k)	3 x	I think my answer was more defensive because I didn't really understand exactly what they were talking about. That WTP question – you need more explanation, you're not actually qualified to say exactly how much, whether it should be twice, three times. I think you can put a value on nature but *not* a value in monetary terms. A value is what we teach our children.
John (age: 55–64, social worker, £40–50k)	100 x	I must admit I discussed that with my wife because I found it a bit difficult to do. I was surprised that I was prepared to spend a bit of money on Pevensey. If someone said OK, now put your money where your mouth is, I would. Paying farmers to do something rather than not doing something must be good.
Daniel (age: 25–34, retail shop manager, £10–15k)	500 x	The way it came across was that it would be for all environmental needs. So I very willingly said fine, increase the taxes, go on, if it's saving the planet. Without all the information you can't get the real answer. Surely it's better to conserve it now than pay the cost of trying to replace it in future years?

Table 7.5   Summary of individual visitors' comments on the Pevensey CV survey (from the visitors debriefing and focus group discussion)

Group member	WTP figure	Reactions to debriefing question and comments in the focus group discussion
Betty (age: 65–74; retired nurse; £10–15k)	refused	I think that's an impossible question. How can I possibly say how much when it's in the future? I don't think I have any means of knowing the value. It's very important to me that these areas are preserved but I don't know what the relevant costs are. I would pay the earth if I could afford to. But you can't really value it can you?
Susan (age: 55–64; retired, medical lab technician; £5–7.5k)	2 x	It's difficult to say if what I said I was willing to pay is an accurate figure. It's difficult to quantify and difficult to compare with these other areas that concern me. You can't put a money figure on your values.
Colin (age: 35–44; farmer; Refused to state household income)	2 x	I think the Levels are worth more than I would be willing to pay. Whatever it costs, that is what should be paid. Surely that is what the Marsh is worth? What people would pay per acre for the land. But nature is priceless – there is no upper limit.
Derek (age: 65–74; retired administrator; £7.5–10k)	10 x	The amount you quoted that we were already paying seemed so low that I was quite happy to say I'd pay twenty times that amount because that would be infinitesimal wouldn't it? I've often thought that specific areas such as Pevensey Marshes and things like this – they should have some sort of scheme where they are virtually preserved for all time. I don't think a money value comes into it.
Yvette (age: 55–64; retail shop manager; £20–30k)	20 x	The money I agreed to pay is probably not a good measure of what preserving Pevensey Levels is worth to me. It's what I can perhaps afford bearing in mind all the other charities which I support. And I presumed that there were other people who would pay for other places. How does the government get away with destroying SSSIs? It has to be conserved for what it is.
Norman (age: 35–44; fireman; £15–20k)	100 x	I don't go there that often. If I was going every day I'd pay more. I'd like to see all this area conserved but I can't say 'Well, the Pevensey Levels, it's better than Rye'. You can fill in all the questionnaires you want, and then they say 'Oh yeah that's very interesting' and, choonk, it's gone. At the end of the day there are people so powerful they just overrule everything.
Wilfred (age: 75+; retired civil engineer; £15–20k)	Over 1000 x	I think the money I said is an accurate measure but my problem is I can think of dozens more round the country and I'd dearly love to spend the same amount on each. But I wouldn't have that amount of cash. I looked at it on the basis of what I would pay to preserve Pevensey Levels but the question was a bit nebulous because it didn't give the context. You can't quantify a field full of orchids. This government says to hell with anything that doesn't make money.

Evidence for an economic valuation

The first response given for each person in the tables is the answer they gave when asked directly what their WTP figure meant. Although complicated by the very small amounts involved, some of the participants gave a WTP figure that represented an amount they felt they could afford: Kathy for example, in the Pevensey group; Laura, John and David in the Maidstone group; Derek, Yvette, Norman and Wilfred in the focus group. In these replies, individuals are assessing the various demands on their income, and then making a judgement about how much nature on the Levels is worth to them and, therefore, how much they would be willing to pay for the Wildlife Enhancement Scheme. You will note how variable the amounts they bid actually are; and both Daniel's 500 and Wilfred's 1000 times the few pence now paid through national taxes would be disregarded as extreme bids. In Daniel's case, that is probably the correct decision because, as you will see later, he did not really understand the hypothetical market. Carla, in the Maidstone group, says she gave a higher figure than she really meant, probably to impress the interviewer.

Some of the respondents expressed dissatisfaction with the valuation task itself and responded in different ways. So, Christopher (Pevensey), Ray (Maidstone), and Betty (visitor group) all refused to give a WTP figure. Others, such as Bob (Maidstone) and Keith (Pevensey) did not understand the task: the former said he would pay three times the amount now being paid in taxes for the WES, while Keith bid 1000 times and again has had his WTP figure excluded from the analysis. Others were concerned about the 'payment vehicle', that is, the fact that farmers on the Levels were being paid under the WES to farm in conservation-friendly ways. The pros and cons of this course of action were widely discussed in the groups (see Burgess et al., 1998) with considerable disquiet being expressed by some members of the groups. In these tabulated responses, Malcolm (Pevensey) is more explicit in addressing the issue, even though his WTP was 100 times; while John (Maidstone) preferred to pay farmers to do something (i.e. manage for nature conservation benefit) than do nothing, as under set-aside schemes.

Evidence for non-instrumental values

In a standard CV, the WTP figure is regarded as sufficient to represent the value of an environmental 'commodity' to the individual. But, as you have seen from the tables, no one was prepared to have that figure stand by itself as a fair representation of the value of conserving wildlife on the Levels to them. It is clear that many people in the groups did not feel it was 'proper conduct' to express the value of nature in money. Of the twenty one participants, sixteen expressed hostility to the idea. People objected on the grounds that nature is 'priceless' in the sense that the values that nature embodies – the intrinsic rights of species to exist, the web of life that encompasses all life on earth, the aesthetic beauty of the countryside, the capacity for nature to teach human beings about other ways of living in harmony with it, and the desire to leave a world at least as rich in species as it is now for future generations. Bob, in the Maidstone group, caught the sentiment well: 'I think you can put a value on nature but not a value in monetary terms. A value is what we teach our children'.

Activity 7.6

Re-read the comments in Tables 7.3 to 7.5, this time looking for comments about how powerful interests make their decisions. Can you find evidence that people are contesting the *status quo*? For example, do the individuals express trust in decision makers? Do many people seem to be contesting how conservation decisions are made, and by whom? Do you get a sense that these members of the public feel they have the power to change the way decisions affecting the environment are made?

Comment

What did you find from reviewing the groups' responses? The first thing to say, bearing in mind that the fieldwork was done in 1994, is that there is an element of mistrust and scepticism about the (then) government's commitment to conservation. Some people argued that citizens have no way of knowing that the money would be spent in the right way, especially when the government was allowing nature conservation sites such as Twyford Down to be destroyed for a road-building scheme. And there was resistance, captured in Keith's last comment (and explored more fully in the conversation between members in the Introduction), to the free market, neo-liberal ethos of the Conservative government of the time.

A second element of the argument concerns the transparency (or not) of the CV survey itself. Members of the groups expressed doubt and some uncertainty about the survey, what they had to do in it and why they had to do it. As Christopher said; 'I didn't like the questions. I wasn't happy answering them. I couldn't place what it was really asking'; or Laura in Maidstone who wanted to be asked a straight question ('don't hoodwink us', she said) by the CV questionnaire. The fact that the great majority of respondents still responded is a measure of the general *politesse* of people, and their faith that if experts knock on the door with a questionnaire, then the questions are worthy of being answered!

Alternative approaches to capturing non instrumental values in environmental decisions.

But, perhaps more important than these considerations, is the fundamental spatial issue that Christopher raised during the discussion in the Introduction to this chapter, and that some of the respondents highlight in the tables. At what scale is the decision about place-specific nature sites to be taken? From whom is money to be collected? And how is it to be allocated between different place-specific projects? In all three groups, members recognized that government and experts were needed to set standards and arbitrate between different local claims. The fact that Pevensey Levels might be thought flat, boring and unattractive as a landscape makes it more difficult for outsiders to *see* what value it has, but not necessarily to challenge the expert view that it does, in fact, have value and should be protected. Indeed, one of the reasons individuals gave for not answering the WTP question was that they could not make a judgement about the Levels, when they knew nothing of the claims of other places and schemes.

The members of the groups felt that the best alternative to CVM would be to create opportunities for more participatory forms of decision making. These, it was suggested, should be held at local level where interested citizens could meet together with local experts, farmers, local authority officers, councillors and members of the business community. The aim would be to create sufficient time and space for people to discuss the issues, and arrive at a practical judgement about what should be done. There may not be total agreement between all the members of the forum, but one outcome would be a measure of consensus, based on a much richer understanding of the different interests and positions represented. These local fora could then be linked to similar regional and/or national level decision-making structures. There is no space for us to pursue this issue further here, but it is addressed further by **Blowers and Hinchliffe, 2003**.

Summary

We have covered a lot of material in this section. It will be helpful to summarize in two stages. First:

- The standard CV survey cannot provide much information about how individuals arrived at a WTP figure.
- Some studies suggest that respondents do not respond in an economically rational way to CV surveys of non-use values.
- A range of reasoning processes appear to lie behind WTP figures for non-use values.
- In these circumstances, CV surveys appear to be less reliable than hoped in monetizing values for environmental 'commodities'.
- Qualitative research in the UK has shown that, although respondents try to express their values in a WTP figure, there was deep unease with the task.
- The CV methodology is expert-driven, opaque and confusing for people who would prefer a more open, deliberative and participatory approach to environmental decision making.

Second, Figure 7.13 highlights the range of situations in which CV/CBA can be used with confidence, and those where neither method is adequate for democratic decision making. Figure 7.13 illustrates the latest expert guidance to the European Union on when monetary valuation can be robust and meaningful, and when it is probably less so. In terms of the vertical axis ('scale'), monetary valuation is useful up to a point. Once the ecological system becomes too complex and the timescale becomes too stretched, technical errors may become unacceptable. The horizontal axis distinguishes between instrumental ('commodity-type') values and non-instrumental ('non-use') values. As the figure shows, the more a decision situation engages with social and cultural values, the less appropriate is monetary valuation.

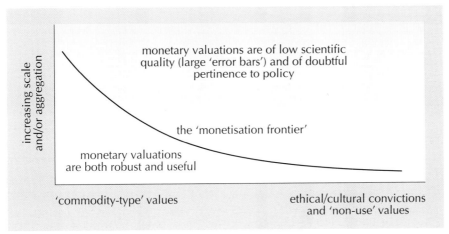

Figure 7.13 The frontier of monetisation.
Source: O'Connor, 2000.

5   Conclusion

The fundamental problem, living in a capitalist-dominated world, is that nature and the environment appear to be highly valued but possess no market to allow those values to be expressed in monetary terms (see Figure 7.14). Substantial progress has been made in the last few years to establish clearly in which decision contexts CV/CBA is appropriate and in which it is less so. The case for CBA in environmental policy appraisal rests on an important assumption: that the market is the appropriate forum through which to articulate values, i.e. individual consumers are as willing to trade in environmental 'goods and services' as they are to trade in other markets. In this way, the previously hidden subsidies that society and the environment has provided to capital will be exposed and redressed.

However, the main technique used to capture the tradable values for nature (the contingent valuation survey) is not universally applicable.

We can now answer the two questions posed at the start of this chapter:

1   How can the different values that people attribute to landscapes, nature and biodiversity be accommodated fairly in decisions about alternative courses of action?

2   How much power do local people have to influence these decisions?

Figure 7.14 The environment may be highly valued, but possesses no market in which its value can be expressed.
Source: cartoon by Neil Bennett.

First, as we have seen, environmental economists have argued strongly that TEV, supported by monetary estimates gained through contingent valuation surveys, is a robust mechanism for ensuring 'economic democracy' in decision making. The discussion has shown that CV (and CBA) is a reductive process that, in the context of non-use values, forces incommensurable values into a monetized sum that does not represent the 'worth' of nature to individual respondents. Individuals do not stand in a consumer relationship to the environment in these cases. Indeed, the extent to which nature and ecosystems can be 'commoditized' (divided into discrete 'goods and services' that can then be traded in a hypothetical market) appears quite limited. The nature of our environmental experience is to be embedded in places; to see landscapes as wholes rather than fragments, and for people to think of themselves as part of a 'web of life'. The ethical values that constitute a culture's code of conduct towards nature are of fundamental importance and, for many people, it is improper to try and reduce that importance to a sum of money. The fact that people do produce numbers to a CV survey is much more a consequence of the individual's trust in the expertise of those asking the questions than a true measure of environmental value. As Schkade and Payne observe: 'People often find CV questions difficult to answer but if you ask someone for a number, they will generally oblige with one, finding inventive and sometimes surprising mechanisms for answering' (Schkade and Payne, 1994, p.103).

Second, we can conclude that trying to force non-instrumental values for nature and environment into a commodity framework runs counter to social experience.

Nature and the environment are meaningful as a public good belonging to everyone. Members of the public expect their elected representatives to make decisions about resources on the basis of scientific, political, economic and social judgements about appropriate standards in the allocation of scarce resources, rather than individual willingness to pay. People know decisions have to be made between different priorities and different courses of action. They are not politically or economically naive. However, both scientific knowledge and local knowledge have a role to play in decision making about locally valued environments. Strategic decisions must be taken by those who can be trusted to make fair judgements and choices between different local claims. The challenge is to construct new institutions within which the usefulness of economics to help make rational choices over limited resources may be complemented by other forms of social intelligence about what should be the important criteria in environmental decision making.

Responses to such challenges are discussed in **Blowers and Hinchliffe, 2003**.

References

Bateman, J. and Turner, R.K. (1993) 'Valuation of the environment, methods and techniques: the contingent valuation method' in Turner, R.K. (ed.) *Sustainable Environmental Economics and Management: Principles and Practice*, London and New York, Belhaven Press.

Blowers, A.T. and Hinchliffe, S.J. (eds) (2003) *Environmental Responses,* **Chichester, John Wiley & Sons/The Open University (Book 4 in this series).**

Boyle, K.J., Desvousges, W.H., Johnson, F.R., Dunford, R.W. and Hudson, S.P. (1994) 'An investigation of part-whole biases in contingent valuation studies', *Journal of Environmental Economics and Management*, vol.27, pp.64–83.

Burgess, J. (2000) 'Situating knowledges, sharing values and reaching collective decisions: the cultural turn in environmental decision-making' in Cook, I., Crouch, D. and Ryan, J. (eds) *Cultural Turns/Geographical Turns*, Harlow, Prentice Hall, pp.273–87.

Burgess, J., Clark, J. and Harrison, C.M. (1998) 'Respondents' evaluations of a contingent valuation survey: a case study based on an economic valuation of the wildlife enhancement scheme, Pevensey Levels in East Sussex', *Area*, vol.30, no.1, pp.19–27.

Clark, J., Burgess, J. and Harrison, C.M. (2000) 'I struggled with this money business: respondents' perspectives on contingent valuation', *Ecological Economics*, vol.33.

Department of the Environment (DoE) (1991) *Policy Appraisal and the Environment: a Guide for Government Departments,* London, HMSO.

Foster, J. (ed.) (1997) *Valuing Nature: Economics, Ethics and the Environment*, London, Routledge.

Hanley, N. (2001) 'Cost–benefit analysis and environmental policy-making', *Environment and Planning C*, vol.19, pp.103–118.

Harrison, C.M. and Burgess, J. (1994) 'Social constructions of nature: a case study of the conflicts over Rainham Marshes SSSI', *Transactions, Institute of British Geographers*, NS19, vol.3, pp.291–310.

Hinchliffe, S.J. and Belshaw, C.D. (2003) 'Who cares? Values, power and action in environmental contests' in Hinchliffe, S.J., Blowers, A.T. and Freeland, J.R. (eds).

Hinchliffe, S.J., Blowers, A.T. and Freeland, J.R. (eds) (2003) *Understanding Environmental Issues,* **Chichester, John Wiley & Sons/The Open University (Book 1 in this series).**

O'Brien, M. and Guerrier, Y. (1995) 'Values and the environment: an introduction' in Guerrier, Y., Alexander, N., Chase, J. and O'Brien, M. (eds) *Values and the Environment: a Social Science Perspective*, Chichester, John Wiley & Sons.

O'Connor, M. (2000) 'Natural Capital' *Environmental Valuation in Europe, Policy Research Brief 3*, Cambridge University, Cambridge Research for the Environment.

Pearce, D.W. (1993) *Economic Values and the Natural World*, London, Earthscan.

Pearce, D.W. (1998) 'Cost–benefit analysis and environmental policy', *Oxford Review of Economic Policy*, vol.14, pp.84–100.

Schkade, D.A. and Payne, J.W. (1994) 'How people respond to contingent valuation questions: a verbal protocol analysis of willingness to pay for an environmental regulation', *Journal of Environmental Economics and Management*, vol.26, 88–109.

Willis, K.G., Garrod, G.D., Benson, J.F. and Carter, M. (1996) 'Benefits and costs of the wildlife enhancement scheme: a case study of the Pevensey Levels', *Journal of Environmental Planning and Management*, vol.39, pp.387–401.

Chapter Eight

Making environment news

by Joe Smith

Contents

1   Introduction

Where do you get news about the environmental debates and conflicts from, and what does it *do* to you? What can you do with it? Who has power over what gets into the news and what is kept out, and how much power do audiences have to respond to the news they consume?

These questions are going to help to structure this chapter and also act as ways of making the book's key questions – particularly 'How are environmental understandings produced?' and 'Can environmental inequalities be reduced?' – relevant to our concerns here. Section 2 gives an overview of why the news media count in environmental debate and conflict. It tells the story of the close relationship between news media and environmentalism in recent years. Section 3 takes one particularly revealing historical case and goes through it in some detail to show how and why environmental debates and conflicts arise. It looks at the story of the Greenpeace campaign in the mid 1990s against the sea dumping of a redundant oil storage facility, the Brent Spar. This case caused all parties – environmental non-governmental organizations (NGOs), government, business and the media itself – to take a long hard look at how the media and environmental issues interact. The issues raised by this case study are relevant to many environmental topics you might pick up in the news today. It clearly demonstrates the news media's pursuit of three things above all as it constructs stories: conflict, event and personality. This section gives you tools to consider the role of the media in the emergence of environmental debates and conflicts.

Section 4 moves on to look at how our capacity to act on what we see in the media is changing. Newspapers, TV, radio and the web all connect their audiences to

Figure 8.1 What role do journalists play in environmental debates and conflicts?

people and places that most of us cannot link to in any other way. They also allow us to keep a close eye on decisions made by politicians and businesspeople all over the world. The news media keep their readers, viewers and listeners up to date with new research about interactions between people and their environment, locally and globally; they can bring the reality of environmental change into our living-rooms and workplaces. It is where most people get their information about climate change, biodiversity loss or the latest scientific or political controversy. Social changes, economic globalization and the constant evolution of media and communication technologies are giving audiences new kinds of understanding and new kinds of power. We all interpret news media in our own way, and in many instances act on and make news, whether in small ways or large. Hence, although the daily bombardment of news texts, sounds and images can leave people feeling powerless and confused, many are also finding new ways to act on information received.

Of course, there's more to the media than news: there is advertising, film, drama, sitcoms, magazines and so on, and these can all be relevant to environmental issues too. However, news has played – and continues to play – a big part in the evolution of debates and conflicts about environmental issues in the last thirty years, and so this chapter focuses on news media in print, on TV, radio and the web.

Activity 8.1

Think of an environmental issue that you first became aware of via news media. Now look at the questions in the column headings on the left in the table below. Make notes in answer to these questions as to which media you recall covering the issues. What sticks in your mind – images, quotes or headlines? In the boxes on the right I've noted down an example of one environment report that sticks in my mind from when I was first finding out about environmental issues.

Table 8.1

What issue, which media and when?	TV reports on ozone depletion in the mid 1980s. Newspaper reports too – but I don't remember them as clearly. Industrial chemicals – CFCs – targeted as the main source of the problem.
What places and people featured in the report?	Scientists had made measurements in Antarctica, and were interviewed in labs.
What images stick in your mind?	A very real hole! (but on a computer screen; multicoloured mapping images).
Why did it matter?	Unknown long-term consequences were mentioned, but threats to human health were a memorable part of the story (skin cancer).
Was anything missing?	How much did it matter to me in the UK? Who were the main culprits? Could we do enough about it? Could I personally do anything about it?

Comment

Doing this activity got me thinking about the way in which the media work (and work on me). There are some general points that came into my mind when looking back at the ozone depletion coverage:

- Media coverage, public environmental concern and policy change are linked: these reports on ozone depletion contributed to changes in public opinion and government policy. They are one of the things that got me and many others interested in environmental issues.
- The media build up stories out of contributions from a range of sources, ideally authoritative figures, such as scientists, government ministers or campaigners from prominent environmental groups.
- Images can be very important to TV and print media: they can act as evidence and stick in the mind (and a story with a good strong picture is more likely to make it onto the news).
- News media reports are typically brief and incomplete; they rarely have the space to go into issues in depth.

2   Environmentalism and the media

Are you an environmentalist, and did the media make you one? Well, in a broad sense, anyone studying this course is likely to have a strong interest in environmental issues. It is also likely that a TV report, a newspaper article or a radio story played a part in sparking that interest.

Environmentalism is just one of a number of social movements (others included feminism, and the peace and civil rights movements) that emerged in the 1960s and 1970s and represented a new kind of political activity. These movements weren't based on class or regional lines, but arose out of the fact that groups of people with shared concerns organized to campaign for change. Coverage in the news media mattered to all these new social movements; it was one of the main battlegrounds where they could contest existing mainstream opinion, and it also helped to recruit and sustain supporters.

Environmentalism and the news media have lived together in a difficult relationship for over thirty years. The way NGOs work, the way they choose priorities, and the degree to which the public think of the environmental movement as behaving legitimately, have all been shaped by the treatment of stories about the environment in newspapers and on TV and radio. You could say that to some extent the environmental movement has been made by the media. What pictures come into your mind when you think of environmentalists? For me, it's Greenpeace, with its rubber boats and whales; Friends of the Earth, with smokestacks; and the Worldwide Fund for Nature (WWF) with photogenic endangered mammals. Even when the organizations themselves believe that they've moved on to new topics and ways of working, these images can persist. This section will take you through a brief account of how environmentalism and news media have fed off each other over the last three decades or so, and explore how environmental issues made it onto the political agenda.

2.1 The media made environmentalism

News media have been important to the development of environmentalism, so it is important to understand something of the way in which the news media work. This subsection will therefore outline newsroom practice and culture, and in so doing will show why the media have sometimes been as much an obstacle as an aid in environmental debate and decision-making.

Events make news, underlying forces don't

Where news media might give a flood of coverage to forest fires or an oil spill in one week, they may miss equally significant events the next, simply because there is a royalty/popstar/sex/football/terrorist event that seems more news-worthy. We shall look more closely in this section at what makes something newsworthy. It is sufficient at this point to note that many environmentalists and commentators argue that if global environmental change issues really are the greatest long-term challenges facing contemporary societies then they should be covered in more depth and with greater consistency.

The occasionally enthusiastic – but frequently erratic and often shallow – coverage of environmental issues is not new, and indeed has always been a feature of the relationship between environmentalism and the media. The discovery of mercury poisoning in Minimata Bay in Japan in 1959, the Torrey Canyon oil tanker spill in 1967 just off the Cornish coast in the UK (Figure 8.2), the leak of dioxins at Seveso in Italy in 1976, and the partial melt-down of a nuclear reactor at Three Mile Island in the US in 1979 were all graphic disasters that received wide coverage in the media across the developed world.

This coverage generated widespread public concern that in turn fed consistent growth in the numbers and influence of environmental NGOs. The potency of a stock of images of environmental destruction was not lost on environmentalists, journalists or even political leaders. Some at the time claimed that British Prime Minister Harold Wilson had decided to attempt to bomb the leaking Torrey Canyon in order to ignite the oil slick, not because it would be effective but because it would appear decisive on the evening news bulletins. These events were newsworthy precisely because they were dramatic, involved victims (people's health or livelihoods, or oil-drenched beaches and seabirds) and could be made visually arresting. Such stories enjoy a brief spate of interest, yet the media's lens always moves on to the next newsworthy item. Environmentalists of the period constantly bemoaned the media's insistence

Figure 8.2 The Torrey Canyon breaking up: images such as these helped make environmentalism.

on reporting these as disconnected media events rather than as potential symptoms of a malaise inherent in western economic and political systems.

The late 1970s and early 1980s saw debate and concern about global environmental issues intensify. Irreversible loss of habitats and species and, by the mid to late 1980s, climate change became the central concerns of the environmental movement. The news media responded to these stories much as they would to any other: they looked for conflict, event or personality, and moved on as soon as other fresh stories offered these key components. Hence the media continued to perform an, at best, ambiguous role in public understanding and debate. While there was some coverage of issues surrounding biological diversity, it was erratic and relied on the unpredictable and shifting gaze of journalists. This raises an important point about the media's claim to have a broad spread of coverage. News editors like to think of themselves as acting as a kind of net, harvesting news in an even-handed manner from around the world, and making skilled judgements about newsworthiness. Of course, nets have holes, and the problem with modern news agencies that are run on tight budgets and deadlines is that the holes in the net are far too big to allow consistently balanced coverage (Tuchman, 1978, pp.21–3).

Hence similar habitats, facing similar destructive events (whether resulting from human or non-human causes), do not receive the same coverage. A comparison of news coverage of forest fires in the Amazon and in Indonesia demonstrates how the culture and practice of journalism shape and limit wider understanding of biodiversity issues (Harrabin, 2000, pp.59–61). In this comparison, holes in the news net meant that an editor (the senior decision-maker in any news outlet) was willing to take a piece from a correspondent on hand in Kuala Lumpur but would

Figure 8.3 Destroying illegally poached animal skins. A strong 'picture story' – environment stories about the non-human world can get into the news, but only if the subjects pull at our heart strings.

not break their budget to fly a reporter to the Brazilian jungle for a single story. Similarly, received editorial opinion judges that stories about humans (and economies) in highly populated areas affected by palls of smoke from a neighbouring forest inferno are more interesting to people than fires in a sparsely populated rainforest. This builds a bias against coverage of the non-human world into news reporting.

Activity 8.2

What is it that gets a particular issue – such as the Indonesian forest fires – into the news? Make a few notes on how you imagine a particular story is selected to make it onto a newspaper front page or into a leading TV news feature.

Comment

Editors – the key gatekeepers who dictate what will or won't get coverage – are informed by a net of news-gathering. They will listen to their own journalists, and keep an eye on other news outlets and newswire services (such as Associated Press and Reuters). Occasionally a story may emerge when an individual news output breaks a new story – an exclusive. More often there is a shared sense amongst many editors that something has become news. However, 'this net has holes' news may not be new, and there may be events that are newsworthy but will fail to come under the media's lens for practical reasons (cost, location of journalists), or because a particular place or theme does not resonate with the editor's sense of their audience's values or concerns. You could take the fishing-net comparison one step further, and see many environment stories as 'throw back': fish that are too small to sell and so are just thrown back into the sea.

News values and the news-gathering net

News-gathering is a process that relies on thousands of individual judgements made every minute of every day. This often leads to distortions in the coverage of complex or long-term issues. Biodiversity, like food, poverty and land use problems, has long been particularly vulnerable to under-reporting or unbalanced messages. Hence coverage of threats to 'charismatic' mammals such as tigers, whales, elephants or seals can be achieved because they can on occasion fulfil particular news values, while more systematic coverage of links between environmental change and the fulfilment of human needs and activities would never make it into the news. Much hangs on whether a particular issue satisfies an editor's sense of news values. **News values** are the core of an editor's or   news values
journalist's professional persona: they are the characteristics that dictate whether a story will run on the front page or head straight for the bin. News values are difficult to pin down and you would be hard pushed to hear clear evidence of their presence in a newsroom. It is important, however, to recognize that editors don't have a free hand, and there are other sorts of power at play, including:

- opinion from a proprietor (owner) who may have other economic interests that might be threatened by a particular editorial line, or may want a particular political perspective to be represented in their paper or broadcast channels;

- information from marketing departments about whether viewer/reader/listener figures are going up or down;
- figures from advertising departments about particular features that will help to sell space or airtime: tourism, gardening and motoring stories can help revenues; too many difficult science, ethical or environmental stories can damage them.

Activity 8.3

Take a few minutes to put together some notes in your own words on how the dominance of news values in media decision-making makes editors powerful in environmental debates.

Comment

When you switch on the TV news or pick up a newspaper it may seem reasonable to think that you are being presented with a fairly balanced survey of what is new and important in the world. However, it's difficult to escape the fact that you are getting a heavily edited and subjective view of recent events. News values are far from being neutral and transparent: they are an unwritten professional code and have a strong influence over the construction of the daily news agenda. Hence the editorial team's own views about what is and isn't news can shape what audiences perceive to be important in the world. Editors are powerful in shaping which issues are presented as worthy for public debate and action, and which are excluded. If they perceive a story to be un-newsworthy, either because they feel it has already been covered, or because there isn't any interesting conflict to portray, then they will not cover it. Hence their power to *exclude* stories can be as important as their power to *include* them. As climate change science and policy became less contentious in the mid 1990s – but more integrated with other policy sectors and in a sense more important – it became less newsworthy.

2.2 Personality plus politics equals news

The media were central players in a key development in environmental politics, wherein environmental issues became considered mainstream concerns in the late 1980s and early 1990s. Opinion poll evidence, steady growth in the membership of environmental organizations and record votes for green parties in many parts of Europe all pointed to changes in the underlying values of the public in European countries.

Politicians were starting to respond. In the wake of persistent lobbying by senior advisors and civil servants, Margaret Thatcher made an unexpected and surprising speech on climate change that was seen as giving the issue credibility at international political level. Although the media showed no interest the first time she gave the speech, when she presented the same material in the glare of the 1988 Conservative Party conference lights a few weeks later, climate change became mainstream political news for the first time in the UK. An event such as this gave the media something to report. It gave them a peg – a moment in time, a voice and a picture – with which to represent the proposal that human activities were changing the world's climate in potentially hazardous ways. The scientific and policy significance of climate change was not different before or after Thatcher's speech, but these important issues did not represent news. But a major statement by a Prime Minister – made at one of the ritual high points of political reporting – put climate change on the map in the UK. From this point on, mainstream politicians sought to create a sense that 'we're all greens now', and that there was consensus on the need for action.

Figure 8.4 At the Conservative Party conference, 1988, Margaret Thatcher made climate change into news in the UK.

As with Thatcher's climate change speech, the United Nations Conference on Environment and Development (UNCED), held in Rio de Janeiro in 1992, offered journalists a number of ingredients. It represented the largest gathering of world leaders for a conference of this kind, and as such could be reported as mainstream politics. UNCED was also a site of political conflicts that the media had not met before. The NGOs were present in force, and were organizing and campaigning in novel ways, including high levels of international coordination, and the nurturing of new alliances. They effectively held a parallel conference, and their presence had to be recognized by the official meeting. Using a strategy that has been deployed ever since, NGOs from the developed world allied with indigenous people's groups in a way which provided news media with (often photogenic) witnesses and/or victims of unsustainable growth that allowed the media to make difficult issues understandable to their audience. Media coverage helped to cement UNCED as a key moment in environmental politics.

Waves of media interest

The layers of complexity and uncertainty that often accompany environmental issues suffer from the tidal flow of waves of media interest and coverage. News editors are acutely conscious of not wanting to return to stories that have been covered before, and feel that their audiences expect them to continually move on

Figure 8.5 Indigenous campaigners at UN meetings put a face to the impacts of climate change.

Figure 8.6 Editors' choices about what to cover in an event often seem to say: 'You want the media's attention? Then break a window.'

to break news about what's new in the world. Hence, enthusiasm for reporting environmental stories around 1992 did soon begin to subside. Ironically, as the politics of global environmental change issues began to solidify at local, national and global levels, with most mainstream politicians seeking to claim green credentials, such stories became less news-worthy. The number of specialist environment correspondents fell away and 'environment' became a form of niche reporting, subject to occasional (and often unpredictable) explosions of interest. The rise of new waves of media interest are generated by scientific breakthroughs, disasters or international environmental conferences such as the World Summit on Sustainable Development in 2002, and evidence of shifts in public (and therefore audience) opinion. However, the issues surrounding climate change and biodiversity loss, energy and agricultural technologies, water or trade of the sort covered in previous chapters in this book are complex and often fraught with uncertainties. As one commentator said, these stories 'don't break, they ooze', making it difficult for journalists to fit them into established and rigid formats. Environmentalists who criticize the media for failing to sustain continued interest have themselves failed to understand how the news media measure their own performance.

Within a decade the reporting of protest and NGO campaigning around international issues (whether on climate change, poverty or against trends in world trade) had become an established strand in media reporting. Indeed coverage of protest has in many cases focused on moments of conflict, to the virtual exclusion of discussion of the issues behind it. The difficulty that the news media have in going beyond a simple representation of two sides in a conflict, or giving a splash of coverage to an event, or ignoring an issue because it lacks strong voices or personalities, is something that Section 3 will explore in more detail. It looks closely at the news media's epidemic of interest in the fate of the Brent Spar oil storage facility.

Activity 8.4

For the next few days, keep track of established news sources – papers, television or radio – and note down examples of environmental stories that are being covered, be they local, national, global or a mix. Many of these may fit into conflicts or debates that you have met already in this book, concerning, for example, food and life-sciences technologies, energy, water, biodiversity or landscape conservation and loss. Think about the following questions:

- Why do these stories fit with an editor's sense of news values?
- Did you sense that there was anything missing from the coverage?
- Where can you get information and commentary that overcome the constraints of news media?

Hint: if you had been working on the last question in 1990 you would have been guided to a public library's journal shelves. Although this remains a very good resource, you will now find a flood of material on the web, and that's where I suggest you go. You should restrict yourself to spending no more than an hour on this activity in total. Taking your search of web materials a step further, you might want to go to the press releases from some news sources, which can be found on the websites of Greenpeace or Friends of the Earth, or an NGO-oriented news site, or UN-related environment bodies. You are likely to find that there are many instances where important issues don't make it onto the news agenda at all.

Comment

The week I was drafting this activity an event occurred that went near to the top of the list of news stories. The Larsen B Antarctic ice shelf fell into the sea, and widespread news reporting represented this as evidence of climate change caused by human actions. This was a palpable event that had been predicted by some Antarctic scientists. It was dramatic in scale and offered impressive pictures (see Figure 8.7). Some of the essential requirements of news values were met, and there were images quite freely and cheaply available to newsrooms. I was interested to find that family, neighbours and – yes – a taxi driver wanted to talk about both this piece of news, and climate change in general. The news made a difference to public debate. However this news had come out in a relatively quiet week. If the story had broken just a little later, and had then coincided with the Queen Mother's death and funeral, the press in the UK would not have covered it to anything like the same degree.

Figure 8.7 The satellite photo image and maps combine to bring distant impacts of climate change home to us: the break-up of the Larsen B ice shelf brings home the effect of global warming.

In the context of a few hundred words of newspaper article, or a couple of minutes of a news item, it is enormously difficult for journalists to link to causes of climate change. They might fit in a reference to how the burning of fossil fuels produces CO_2 – 'the most important global warming pollutant' – but it is much more difficult to find space to engage their audience in discussion of household energy use, growth in lorry and car traffic and so on. The communications people in environmental NGOs and UN and EU environment bodies give a lot of energy to trying to extend climate change coverage into these kinds of questions about production and consumption. However, they rarely succeed, and it is necessary to go to specialist or NGO sources to get more depth on an issue.

Summary

In this section you have:

- gained an overview of the relationship between news media and the development of environmentalism over a thirty-year period, with particular reference to the fluctuating waves of media interest;
- been introduced to some of the features that news decision-makers look for – conflict, event and personality – and how the presence or absence of these can affect whether or not environmental stories are covered;
- noted the nature of news-gathering as an imperfect net of coverage (with some significant holes that impact on coverage of difficult issues, distant places and non-human stories);
- considered the centrality of news values in editorial decision-making about environment-related stories, and reflected on them as a source of power (the power both to include and to exclude issues).

3   Making claims: the Brent Spar saga

A handful of good-hearted chaps in a lilo taking on the entire Western economy, and with it, the biggest piece of litter in the world: the Greenpeace war on the abandoned 'Spar was an incident crammed with dramatic polarities of the most unsubtle kind.

(Simon Barnes, *The Times Review of the Year*, 1995)

At first glance the Brent Spar is an unlikely media star. Yet in 1995 this redundant and rusting North Sea oil installation became one of the best-known pieces of engineering in Europe. The conflict about how it should be disposed of is a way of summarizing debates about just what 'the politics of sustainable development' might mean in practical terms. All this was the result of Greenpeace's decision to campaign around the 'Spar and, much later, the European media's decision to pick it up as a story.

This case shows how environmental conflicts emerge, how they are represented in the media, and how effectively debates about solutions are represented to the public. In this section we focus on how claims about the 'Spar were deployed by media sources, primarily Greenpeace, Shell and the British government.

Most of the public's experience of the conflict was mediated through television and print journalism. The story satisfied news values by offering event, conflict and personality. As is often the case with Greenpeace, the organization created a spectacular news event, distilled around the question of whether this oil storage platform, the first of approximately 400 North Sea installations to be decommissioned, should be dumped at sea. The conflict was driven by the more general issue of how industries should act with regard to their waste, whatever its nature and wherever it may be. Shell believed its sea dumping route was economically the best option, and not environmentally damaging. Greenpeace viewed this as a precedent-setting case. Although they raised issues about toxic wastes contained in the 'Spar, and questions about the impacts on aquatic life of dumping it, they were most concerned to have the principle established that

Figure 8.8 Such images covered TV news and front pages, making the 'Spar one of the top news stories of summer 1995 throughout Europe.

industry should re-use and recycle their waste as carefully as the most conscientious householder.

In addition to the conflict between the main players – Greenpeace versus Shell and the British government (who had staked their political credibility on backing Shell's proposal to dump it at sea) – the story also offered a stock of compelling images. In fact the dramatic course of events surprised all players, including Greenpeace who had simply not anticipated the storm of media interest and public debate. Two questions need to be answered: why was this such a milestone in environmental politics, and why did this campaign take off so dramatically in the media?

Box 8.1

The Brent Spar had been in use as a storage tank for crude oil since the mid 1970s in the North Sea. It was decommissioned in 1991 and Shell proposed to dump it in the Atlantic Ocean. The company and UK government pointed to an environmental analysis of this option to demonstrate its acceptability. However, Greenpeace used this – the first dumping of a North Sea oil installation – as a high-profile campaign target. They occupied the 'Spar, and attracted huge media attention which in turn sparked a widespread boycott of Shell products across Europe. The company was forced to drop its plans in 1995 and went on to dismantle the structure and by 1999 it had been recycled as part of a ferry terminal in Norway.

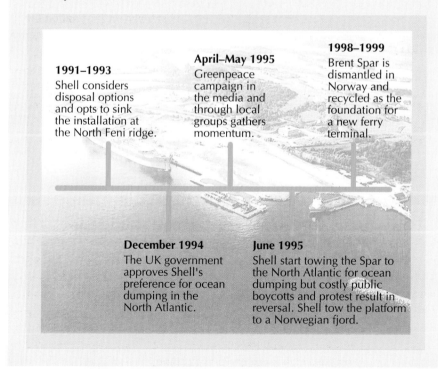

1991–1993
Shell considers disposal options and opts to sink the installation at the North Feni ridge.

April–May 1995
Greenpeace campaign in the media and through local groups gathers momentum.

1998–1999
Brent Spar is dismantled in Norway and recycled as the foundation for a new ferry terminal.

December 1994
The UK government approves Shell's preference for ocean dumping in the North Atlantic.

June 1995
Shell start towing the Spar to the North Atlantic for ocean dumping but costly public boycotts and protest result in reversal. Shell tow the platform to a Norwegian fjord.

3.1 The cast assemble

The early 1990s was a period of consolidation for developed world environmental NGOs. The UNCED conference had promoted 'round-table' processes or '**stakeholder** dialogues' between business, government and NGOs as a way of making progress towards environmentally sustainable development. This approach carried potential costs. It threatened to make NGO campaigns less reportable (less 'event, conflict and personality' to satisfy news values), stretched their limited resources and engaged them in lengthy dialogues that relied a lot on trust. The Brent Spar allowed Greenpeace to remind other players what they believed the stakes around environmental issues to be. The other stakeholders had become adept at deploying the flexible concept of sustainable development. They were producing glossy brochures that often mimicked the images and sentiments of the NGOs' materials. However, the concept of sustainability had made little perceptible difference to government or business practices. Chris Rose, campaign director of Greenpeace at the time, explained the rationale for the campaign thus:

> Here in the northern North Sea was the government slipping through the gate at the end of the national garden, and conniving with Shell to dump the 'Spar at sea rather than reuse or recycle it – behaving for all the world like a couple of bad neighbours who creep out at night and craftily wheel a rusty old car into the village pond.
>
> (Rose, 1998, p.9)

The campaign was only one step in 15 years of activity on international and domestic ocean policy, including work against dumping at sea of radioactive and industrial waste (see Chapter 3), decommissioned nuclear submarines, and sewage sludge. Most of Greenpeace's campaigning had gone on within the slow process of international meetings, with the exception of flurries of interest around some of their campaigns, notably on radioactive waste dumping, earning relatively little media attention. Brent Spar was important to Greenpeace as a precedent: it would shape future decisions on all the North Sea installations. Yet, it also offered an opportunity to draw media attention to business performance on sustainable development.

Shell and the British government did not anticipate trouble over the dumping of the 'Spar. They were satisfied that legal obligations and best practice had been followed. Their claims rested from the start on a firm assertion that the scientific and technical data all suggested that they were following the **Best Practicable Environmental Option (BPEO)**, that is one that seeks to enshrine both good environmental and economic sense in decision-making. They were confident of the rigour of expert risk assessment, and were attracted to the fact that dumping was the most economic solution, both for the company and the UK Treasury. Hence business and government players were caught out by the storm of media interest which broke over Shell's bemused managers.

Stakeholder is now a common term in environmental policy debate and means prominent or relevant actors, i.e. NGOs, government, business and academia, who have a legitimate claim to be involved in debate.

Best Practicable Environmental Option (BPEO)

○   In hindsight it was clear why this became a big story but, at the time, media people could be forgiven for thinking it was just another pressure group stunt. What were the ingredients that pushed the Brent Spar into the spotlight?

●   This episode had: conflict – between Shell and Greenpeace; event/drama – will they or won't they decide to dump at sea?; and personalities and images – this was a great picture story.

3.2 Images in the media

The discussion of the Brent Spar case has so far sketched out the story. There has been less discussion of the images used, although they were clearly significant. Indeed images may have worked more effectively than the spoken or written reportage to represent the issues that the public were responding to when they appeared to support Greenpeace. Consider some of the images that were used to illustrate the story in the news, such as Figures 8.8 and 8.10. At the most immediate level, images such as these in print, television and, increasingly, web-based media might serve to give context, visual interest and information that is difficult to convey in speech or written text. They can suggest the location, scale and condition of the 'Spar. But picture editors on newspapers and TV news editors are offering much more than descriptive information with strong pictures like these. The precarious Greenpeace rubber boats and protestors clinging to the 'Spar, and the opposition with its large ships, water cannon and helicopter surveillance, all tell a particular version of the story in an attention-grabbing second. Greenpeace and other NGO campaigners are highly sophisticated in their construction and use of images. In this case the images served to summarize the story and to engage the viewers' and readers' interest.

However, studies of the mass media argue that something more significant goes on when we consume images. Whether it is news media, advertising or drama, we are faced with a steady flow of images in any day. They aren't just important in aiding our understanding: they also allow us, in a sense, to participate in distant events and processes (Urry, 2000, p.180). Clearly, images were important in both representing and motivating public feeling about the Brent Spar episode, and were working at a different level from the statements made by both sides and by independent commentators.

○   What role do images play in news media?

●   In addition to giving visual interest, and illustrating particular points in texts, images can do other work. They can serve to distil or summarize textual or spoken material. Audiences and readers can become engaged in – in a sense participate in – events that are distant in space or time. They can link our conscious debate about an issue with more personal, subjective responses. Images often work at a different level to texts and the spoken word. This is a key insight when it comes to looking at how environmental conflicts have been represented in the range of modern media.

Faced with an accelerating storm of media interest and public demonstrations, including boycotts of petrol stations throughout continental Europe, Shell announced on 20 June 1995 that it would not dump the 'Spar at sea. Centre-left papers, which had been broadly supportive of Greenpeace, proclaimed a 'people's victory': the *Daily Mirror*'s front page headline 'Glad oil over' (see Figure 8.9) was supported by an editorial which proclaimed a 'Victory for the People', and went on: 'Shell's decision not to sink the Brent Spar oil rig is a fantastic victory ... A victory for Greenpeace, a victory for *The Mirror* and other newspapers which

Figure 8.9 Papers that had sided with Greenpeace were quick to share in the victory: front page of the *Daily Mirror,* 21 June 1996.

campaigned against scuppering the platform. But most of all it is a victory for the people. The people whose boycott of Shell forced it to back down.' Clearly some of the newspapers viewed themselves as components in the success of Greenpeace's campaign.

The Financial Times took a cooler tone, concluding that: '[although] the widening consumer boycotts were not expensive in money terms, they were doing untold damage to Shell's public image. Even the German police were refusing to use Shell petrol' ('Company struggles to accept disaster', 21 June 1995). The right of centre press in Britain framed the story in terms of the negative political implications for the UK government. It also portrayed Shell as having 'lost its nerve'.

Figure 8.10 Boycott of Shell in Bonn, Germany: protesters wanted to express their feelings through a boycott, not argue about the status of the scientific debate.

3.3 'We were spun' say the media

Most of the media, then, were presenting the Brent Spar story much as a David versus Goliath conflict, with the plucky and virtuous underdogs bringing a much more powerful adversary to its knees. And it might well have ended there in news terms, except that Greenpeace discovered they had made errors in sampling and estimating the amount of oil remaining in the installation. The implication of this finding was that the consequences of dumping at sea were not as environmentally damaging as they had previously argued.

This revelation kept the story in the news, and is one of the features of the case that makes it so important to any study of media and the environment. Greenpeace decided that they should come clean and announce their error to the press. Even though the mistaken figures had been little reported when the information was released to the UK media on 16 June, and played no evident part in the unfolding consumer boycotts, the announcement was given almost as much prominence as Shell's decision weeks earlier not to dump. Some coverage was forgiving of Greenpeace's apparent exaggeration, suggesting that it is 'Better to blunder than to lie' (*The Independent*, 6 September 1995). However the more widespread reaction was that the environmentalists had 'spun', or manipulated, the media with their emotive message, their effective use of images and communications technology and, as it now turned out, their suspect science claims. Many editors decided to overcompensate for what they felt had been naivety in their handling of the environmentalists. Greenpeace were now portrayed as at best careless and at worst purposefully deceitful. Much of the coverage that followed seemed to move them from players in legitimate controversy and debate, and back into a category of media outsiders that they had inhabited years before when environmental issues had not been considered a mainstream concern. Their status as a source changed.

3.4 According to whom? ... Sources and source strategies

Media researchers have invested a lot of time in discussion of sources. So, where did this story come from? Greenpeace were the primary source of the Brent Spar story, as well as of many of its abiding images. However, the NGOs are not the only actors in environmental policy debate. Government officials and press officers, politicians, business interests and independent experts all have the potential to act as news sources. Unwritten codes of conduct about how to behave in relation to sources are at the root of journalistic practice. At the same time, people studying the media have shown how sources attempt to ensure that their own contextual interpretation appears in the news story as reported. Though their briefings may appear objective, and meet journalists' demand for facts, they may still be guiding debate in a particular direction.

Sources are key to news-making: they often make stories visible to journalists, and time pressures and demand arising out of the 24-hour delivery of news have seen journalists use materials from sources that are close to being finished stories. Most research in this area suggests that journalists tend to favour government or 'legitimate' sources (nowadays this often includes the better-resourced, research-oriented NGOs). They will keep good relations with such sources, but will be less likely to maintain relationships with some sorts of environmental sources, such as informal groups, distant groups (such as those based in the Third World) or those perceived as radicals. This first stage in the process of news production inevitably leads to distortions in the balance of what audiences receive as news. A large professional industry has developed around creating and sustaining organizations or individuals as sources. Public relations practitioners (in politics often now referred to as spin doctors) try to do half the work of journalists for them by timing the release of carefully packaged information to maximize, or on occasions minimize, coverage. The question of which sources are ruled in or out is of great significance for how environmental conflicts and debates are treated in the media, and can shape what you or I come to think of as an environmental issue.

When news editors got hold of Greenpeace's own revelations about faulty science claims, the NGO lost some of their status as a legitimate source. But this did not lead the media to subject the science produced by government and industry to any corresponding degree of scrutiny. The official or legitimate source was simply given more authority in this framing of the issue. This confidence in 'official' science ignores the fact that assumptions (and errors) were made by all parties – as two government-initiated reports acknowledged (National Environment Research Council, 1996, 1998).Yet the sense amongst some editors that they had been spun by an NGO offended their unwritten code of journalistic ethics. Discussion of the Brent Spar at the Edinburgh Television Festival that autumn threw light on editorial decision-making, and the irritation felt by news media decision-makers towards Greenpeace. David Lloyd of Channel 4 News echoed other editors when he said that:

> On the Brent Spar we were bounced. This matters – we all took great pains to represent Shell's side of the argument. By the time the broadcasters had tried to intervene on the scientific analysis, the story had been spun far, far into Greenpeace's direction ... When we attempted to pull the story back, the pictures provided to us showed plucky helicopters riding a fusillade of water cannons. Try and write the analytical science into that to the advantage of the words.

(quoted in Rose, 1998, p.158)

The fact that many of the potent video images of the 'Spar conflict were provided by Greenpeace became a particular cause for media introspection, encouraged by government and business complaints of uncritical one-sided reporting. Yet in relying so extensively on Greenpeace video news releases (VNRs) for their footage, TV coverage of the 'Spar was simply following an increasingly common practice.

Through the course of the 1990s time and budget pressures in newsrooms has seen this extension of the role of sources in providing stills and moving images, as well as text, interview subjects and ideas, becoming commonplace. Greenpeace's own videos offered editors good picture-led stories, without the cost of speculatively sending camera crews to what may, or may not, have provided a story.

Several key editors felt they had been publicly shown to have failed in their professional judgement, and their irritation at this strongly influenced their treatment of the aftermath of the story in the following months.

Activity 8.5

Use the table below to help you think about the uneven opportunities that different sorts of sources have in shaping news media in the global news centres of major developed world cities. In the left-hand column is a list of different kinds of news sources. The column headings encourage you to get to grips with the strategies that these sources can pursue to get their version of a story into the news. Much of the material you'll need to be able to fill in the table has been covered in one way or another in the chapter, but to fill in some of these you may need to think on your feet.

Table 8.2

Sources of news content	Can they put out a press release that journalists are likely to use?	Can they provide high quality images and / or video?	How likely is it that they will be asked to comment on an issue?
UN body (e.g. UNEP)			
Well-funded developed world NGO (e.g. Greenpeace, WWF or FoE)			
Major global company (e.g. Shell or BP Amoco)			
UK scientist attached to a government-funded lab			
Independent scientist based in India			
Radical grassroots NGO in the developed world (e.g. Reclaim the Streets)			

Comment

It is worth taking time to think about the sorts of individuals and organizations that can grab the attention of journalists in the big global news hubs of Washington, London, Paris and so on. Even a few minutes spent sketching in your responses in the table will reveal the widely varying capacities and/or opportunities to act as news sources.

When you have filled in the table, look at the completed table at the end of the chapter and compare your answers with those given there.

3.5 From environment to sustainability

While the media continued to focus on the implications of Greenpeace's unstable scientific claims – and its artful manipulation of news media – a more significant story had been unfolding around them. The Brent Spar conflict had catalysed some surprising new thinking among government, big business and the environment science and policy community. Yet, lacking in news values, these debates received very little media attention.

Figure 8.11 Europe's biggest piece of re-used waste: the Brent Spar's new life as a quay (in the bottom centre of the photo) in Mekjarvik, Norway. Its new purpose was agreed after extensive negotiation.

Greenpeace's head of campaigns felt that, '[T]he real significance of the Spar is all about corporate responsibility, not about oil or steel or the ocean' (Rose, 1998, p.95). You might be surprised to find that senior figures in Shell had arrived at similar conclusions. They had begun a major rethink weeks before their reversal of the decision to dump, when they were suddenly faced with boycotts of the company in response to their initial decision to dump in mid 1995. Despite Shell's long investment in scenario planning, they hadn't anticipated public reaction to this or other issues that they were faced with in the twelve months that followed.

They simply had not recognized that public views of corporations and their responsibilities were changing fast. Yet once the need for a drastic overhaul was brought home to them, Shell began to pay serious attention.

As well as investing in a major public relations effort to convince the media that its 'science was sound' all along, the company made some significant changes as a result of extensive round-table dialogues between senior management and NGOs. It committed itself to significant increases in renewable energy investments (albeit from a very low baseline). The company decided in April 1998 to pull out of the Global Climate Coalition (GCC), the US-based grouping of oil and car companies that had lobbied against action on climate change. These shifts signalled the beginning of new thinking in some influential parts of the fossil-fuel industry about future core interests. Shell's communications team worked on media outlets and through advertising to recover the brand's previously strong status (see Figure 8.12). At the same time, environmental and social NGOs were ratcheting up their surveillance of corporate performance. The term 'greenwash' was coined to describe empty corporate promises on environment and sustainability.

Figure 8.12 Companies need you to love their brands: an example of Shell's advertising campaign.

Looking back through the account of the Brent Spar's disposal in this section, we can identify two important storylines, of which only the first was covered to any significant degree by the media:

1   A 'David and Goliath' contest in which environmentalists supported by public opinion took on big business and government and won a dramatic victory.

2   An important moment in the development of new thinking about corporate social and environmental responsibility, wherein businesses had to learn fast about sustainable development, about involving a wider circle of opinion in decisions and about how they report on their work to NGOs and the public.

Why was the first of these stories widely covered while the second was neglected? One explanation is that the second isn't news because businesses simply haven't changed very much. But there is a further explanation that we have met earlier in the chapter in relation to climate change: the news media find it very difficult to fit slowly evolving dialogues into their rigid formats. Given that the second storyline about corporate responsibility is arguably of much greater long-term significance for societies, the NGOs' charge that the media are an obstacle to effective progress on environment issues seems to carry some weight. It is worth looking around, however, for other places where these issues can be raised or debated for and by the public. The web is offering one important new location where progress on the second storyline can be assessed. You can go to the websites of major companies, of the big NGOs as well as some specialist corporate 'surveillance' sites (see Figure 8.13) to find out for yourself the state of things in relation to a bundle of issues that news media continue to find difficult to report.

Activity 8.6

The Brent Spar story took several twists and turns. Take on the role of news editor – the boss in any news media office who decides what is and isn't going to make it into that day's news. You can choose whether you are in print, TV or radio, and what sort of audience you're aiming at. Make a few notes that you will pass on to your environment journalist covering the story about what kind of angle you want them to take. Write a few sentences – treating each stage in turn – indicating how newsworthy each stage is, how much prominence as editor you intend to give it, if any, and what kind of angle you want the journalist to pursue. Remember journalism's enthusiasm for event, conflict and personality. The sequence of the story goes as follows:

1   Greenpeace activists board the Brent Spar oil facility; Shell are trying to get them off; Greenpeace are providing some very strong images.

2   Public protests are gathering pace around Europe; this issue seems to resonate with large sections of your readers/viewers/listeners.

3   Shell have backtracked on the dumping decision and embarrassed the already weakened UK government which had been staking political capital on the dumping decision.

4   Soon after their moment of apparent victory Greenpeace have owned up to getting some details of their science wrong; it seems the media were 'spun'.

5 Various companies, governmental and advisory bodies have begun to experiment with new kinds of dialogue with NGOs and the public, including environmental and social reporting and round-table discussions. Some talk of a 'new way of doing business'.

When you have completed your answer, you can compare it with one given at the end of the chapter.

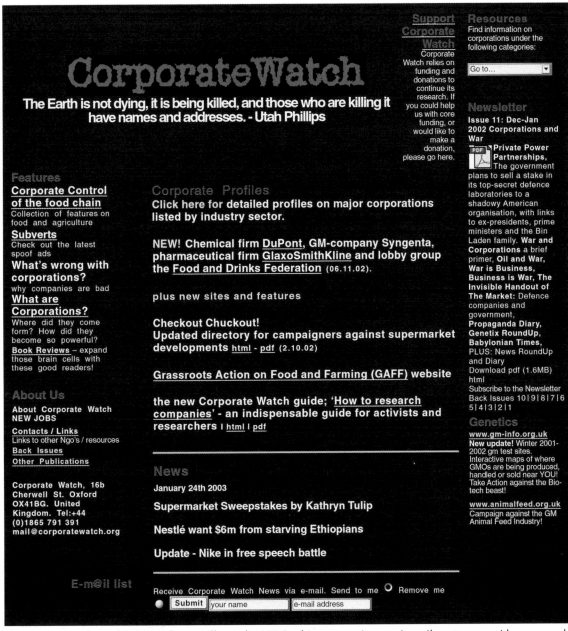

Figure 8.13 With worldwide web surveillance by NGOs, big companies won't easily get away with greenwash: Corporate Watch (http://www.corporatewatch.org/).

Summary

In this section we have found that:

- The media seek stories that provide conflict, event, personality and strong images; a story that offers all of these components has a good chance of gaining sudden and widespread attention.

- Media decision-makers overcome some time and budget limitations by making extensive use of specific sources, including use of images and film provided by environmental NGOs, business and other interests.

- The media select and deploy 'facts' about environmental debates and conflicts according to the journalist's or editor's notion of what the story is about.

- Images play an important role in our experience of the media and work at a different level; they may do a better job than spoken or written text in representing some shifts in environmental debates and conflicts.

- Sources shape outcomes in the production of news stories linked to environmental debates, and sources have varying levels of access and influence depending on skills, time and financial resources, personal links and geographic location.

- Important shifts in thinking and debate, such as those around corporate environmental and social responsibility, are difficult to fit into news formats.

4   Globalization, the media and citizenship

If you were born before 1950 you might recall the excitement around the first live TV broadcasts such as the Coronation of Elizabeth II. If you were born before 1970 you were probably among the millions who gathered around radios and televisions to hear transmissions of the first human voices and pictures from space. Those born in the first decade of the twenty-first century will grow up barely acknowledging the astonishing variety of immediately accessible news and other media. Hundreds of channels of radio, TV and web media join a spread of local, national and international papers and journals to offer up a huge array of news that most people can draw upon. Media production and consumption is one of the fastest-moving aspects of economic and cultural life. There have been important developments in the way in which media, society and politics fit together, even since the Brent Spar saga. This section charts some of the technological, economic and social changes that are shaping these rapid transformations of the media landscape. It also invites you to pick out some of the threats and opportunities these present in terms of effective understanding and debate of environmental issues.

4.1 News: anytime, anywhere, anyhow

Technology changes news

Information and design technologies are opening up new possibilities for both news-gathering and presentation, and highly competitive broadcast news operations are seeking to outdo each other in their use of them. For example, the occurrence of dramatic floods in Europe in 2000/01 (Figure 8.14) saw news organizations working to integrate more simultaneous broadcasts from correspondents than their competitors.

Figure 8.14 We knew where people were getting their feet wet, but did we find out why?

Some argue that these new technologies deliver greater choice and diversity in the news market, and create news 'products' that the public are more likely to consume. At the same time, however, the decision to invest in a display of technically impressive and difficult outside broadcasts may in itself reduce the information value of news. The fact that similar statements are delivered from several different locations ('Ordinary life has been postponed here in ... as the floods ...') may in fact be getting in the way of asking some important questions. Why have the floods had such an impact? Is this an example of what global warming might bring? What are the causes and consequences of climate change? Should developers be building on what have always been known to be flood plains? Even where these questions are addressed, the style of presentation can distract from the content.

Developments in graphics, and the integration of images with texts, are accelerating the amount of information that can be crammed into tight news slots. These advances in the presentation of news items can condense – but also threaten to dilute – television news. One TV news presenter noted that it is now possible to:

> ... package, graphicize and meld five minutes of old TV information into 60 seconds of new TV time. The whizz and bang of such presentation may be enticing but the content reduction is so acute that normal debate is in danger of being degraded to the absurd.
>
> (Jon Snow, 1997)

The shifting economy of news media

Macro-economic trends have also had an impact on the content and character of news media. Economic globalization **(Castree, 2003)** has seen the concentration of media into a diminishing group of global corporations. This has delivered efficiencies of scale and accelerated the uptake of new technologies. However, it also threatens the diversity of content. The process of cross-media ownership and company mergers has become known as media convergence. This has seen energy companies buying TV stations, water companies becoming TV and web giants, and multi-media empires springing out of publishing companies. As well as being multi-media these companies are often transnational – even global – in reach. Ownership of these companies is in some cases in the hands of individual 'media moguls'. The political and cultural power of these organizations and individuals is immense, but difficult to pin down and classify. Nevertheless, there is evidence that the values held by media owners are influencing editorial decisions in newspapers and broadcast outlets that proclaim editorial independence. This is something to bear in mind when you consider how conflicts between environmental and development goals are presented by companies that may be part of a conglomerate that itself has a big environmental impact.

Democratic governments often recognize that diverse editorial opinion and independent comment are fundamental to a well-functioning political system. Yet economic globalization is creating pressure for liberalization that allows companies to grow and compete across borders. As a consequence of this the regulation of media ownership has been recognized as a growing concern.

The chapter started by referring to a dazzling array of news media. Yet in spite of this, media outputs and media content may in fact be becoming less diverse. The relationship between audience research, marketing decisions and media production suggests that news coverage is becoming increasingly indistinguishable. Most of these companies provide news services not out of a sense of duty to civil society, but in the way that restaurants provide ice-cream on the menu: many of their customers want to consume it. Journalists have always aimed to both entertain and inform, and almost all media (with the important exception of public broadcasters) have always been commercial. However, many believe that competition between news outlets has seen the balance shift in favour of entertainment over information, with faster-paced, more graphic and more superficial coverage. Complex long-term issues such as loss of biological diversity or climate change, and the role of humans in these changes, become increasingly less likely to gain effective coverage, even at a time when they are becoming more urgent. Taking British television as a whole, there is evidence of a general trend towards less foreign coverage, fewer environmental programmes, and more travel, wildlife documentaries, light features and shorter pieces.

4.2 What do you *do* with the news?

Environmental understanding and debate are essential if societies' responses to environmental issues are to be appropriate and successful. However, this is not simply achieved by the media generating material that their viewers, readers and listeners soak up like a sponge. Media researchers talk of the way in which audiences can be an important part of making the news. Though editors, journalists, proprietors and sources remain the main players, it is becoming increasingly clear that audiences are much more than passive recipients of news. When you sit down with a newspaper or switch on the TV news, you are doing much more than receiving and filing information: you will interpret and act on news in ways that nobody can accurately predict or control.

Views of the relationship between public audiences and the media are ambiguous. On the one hand, modern news is seen as being potentially disempowering. Sociologists have described the impact of contemporary news as resembling a 'collage of disconnected stories about famines, droughts, slaughters, nuclear accidents and so on (which) intrude and shape everyday life' (Urry, 2000, p.127). It's worth noting how most of Urry's examples link in some way to environment and sustainability issues. Dwelling on two of them for a moment, we can see that brief reports that summarize apparently chance happenings may not just be incomplete, but may get in the way of adequate responses. Famines and droughts are often understood by researchers to be caused by humans, perhaps through local land use or economic changes, or human-induced climate changes, but the reporting of them in the developed world that portrays helpless victims of fate suggests that both the victim and the viewer are powerless to do anything. Similarly, specialists in radioactive waste are likely to point out that nuclear accidents are not chance events, but need to be understood in their complex setting of management systems, political cultures and legal and technical capacities **(Hinchliffe and Blowers, 2003)**. Brief news reports do little to promote public understanding and debate of what should be changed. Hence, distant events enter our lives in reports that last no more than a couple of minutes: it becomes very difficult to understand the set of processes that led to a particular incident. Audiences generally have little or no influence or power to intervene in such events. On the other hand, however, the media coverage can serve to increase our knowledge of causes and consequences. It can also serve to build connections between audiences and people experiencing the effects of often global environmental changes.

Figure 8.15 What will they *do* with what they see?

Some current media researchers argue that, far from being 'dumbed down', citizens are in fact becoming more empowered by the rapidly changing, globalizing media. They suggest that media-literate audiences are able to interpret media texts critically. They have 'the power to resist or reject the message, to contextualise and interpret news in ways beyond the control of any journalist or spin-doctor' (McNair, 1998, p.57). In other words popular culture is not something people receive, but is something we all make. Hence fears that increasingly global news products will see a narrow, Americanized world-view become universal need to be balanced by the fact that many people can unpack media messages, subvert intended meanings and/or create new ones. People are becoming adept at interpreting the influence of sources and other practices of news culture and at making their own sense of what is happening in the world.

Hence biodiversity and climate change stories, rather than fall prey to the effects of recent media changes, may be picked up by audiences well equipped to interpret and act on upon them. I find this an optimistic and incomplete view, and immediately find myself wanting to temper it by thinking about who is and who isn't reading media texts critically. Clearly some people are motivated to talk about and act on global environmental change issues by news media coverage, but I find I want to know more about the – possibly larger – number who are not. Audience research hasn't given us the data we need to make a judgement.

4.3 From receiving news to making it

The evolving web of links between global environmental change, new social movements, including environmental NGOs, and developments in information technologies represents one of the bigger features of change in contemporary environmental politics. Though presence on television news remains their main goal, NGO communications strategies have, since the mid 1990s, moved on from working purely to see their arguments and values represented in mainstream news outlets. Large, well-established groups such as Greenpeace, as well as loosely organized grassroots networks, have been using the Internet to organize supporters, and to campaign directly on issues via newsgroups and e-mail petitions. They have also created independent news resources using cheap digital technologies and the web to ensure that their own account of issues or events can

Figure 8.16 Professionally edited news for activists by activists: the Independent Media Centre has a network of websites across the world (http://www.indymedia.org/).

be accessed widely. These new sources of news are creating a very different, more two-way, relationship between sources and consumers of media outputs.

E-mail newsletters and newsgroups have made communication amongst like-minded individuals cheap and fast, a fact that is helping advance environmental debate and action. Let's take my own e-mail inbox as an example. I'm on a few e-mail newsletter lists that are edited all over the world – in London, Copenhagen, Delhi, Sri Lanka, Toronto, Oaxaca, Barcelona and Norwich. They help me to stay in touch with events locally and around the world, and encourage me to keep an eye on the fact that there are many different ways of thinking about and acting on environmental issues. Any week might deliver polemical articles from an African journalist on deforestation, a review of recent biodiversity losses in parts of India or of climate science from the University of East Anglia. In the same batch of e-mails I'll find news on the role that taxes and regulations are playing in reducing air pollution in Indian cities and a UK local authority's plans to decrease the amount of compostable waste going into landfill sites. The Water Media Network, a newsletter aimed at journalists working in the developing world, might send news of their latest awards or workshops on water issues. The Gary Gallon Newsletter – written by a man I'll probably never meet and know nothing about – may arrive with its occasional unsolicited briefing on business and environment issues in North America. This collage of issues and sources helps to remind me that there's a lot going on in terms of networking, debate and action in the world.

The Internet and e-mail have allowed decentralized political activity to flourish. They permit networking that facilitates large numbers to plan campaigns, organizing both large-scale rally-style protests and more dispersed lobbying by e-mail and phone. They have also speeded up and strengthened the building of coalitions that combine social, environmental and economic concerns (often dubbed 'anti-globalization' by the media) both within a country or region, and over large distances. This is in part because they fit news values considerably better than concerned people protesting politely. Images and storylines of some of these protests in the mainstream media have tended to concentrate on minority instances of violence or damage to property. The independent media routes allow the NGOs to work to correct the balance of these representations at least within their own networks.

The Internet has a significance way beyond the NGOs' uses of it. Some argue that linking up the Internet with existing news media and government promises e-democracy, whereby citizens can find new ways of engaging with elected representatives and state institutions. They promise that this will in turn enhance democracy in ways that the telephone or television could never achieve.

Enthusiasts for e-democracy generally acknowledge that other conditions are necessary for these hopes to come to fruition, including increased general commitment to citizen involvement in decision-making, freedom of access to information and equitable public participation in the technologies themselves. It is interesting to note that journalists are still considered essential as mediators in

Figure 8.17 A cybercafé in Bangalore: computer-mediated communication spans the world – and can change it.

these models of a more vibrant democracy. Such developments hold the potential to democratize not just at local and national government levels, but could also become integral to corporate decision-making where they are trying to engage in openness and action on environmental and social issues. Companies that fear the global scrutiny of NGO networks – such as Shell in the wake of Brent Spar – are seeking to forge direct links with stakeholders via web publications and invitations to participate and comment on progress on contentious issues.

But there are limits to the web. Above all there is the question of access to the Internet: how can rural communities in the developing world, or the urban poor, obtain information about, and participate in, environment and sustainable development debates that affect them? What are the literacy rates? What level of affordable, reliable access to the Internet exists? Who writes, edits and controls the sources of information? To take just one example, 80 per cent of the material about Africa on the web originates outside the continent, and access to the Internet (phone lines and computer hardware) is very unevenly spread, both socially and geographically (Brooks Johnson, 1998). Governments want fuller participation in politics via the web, but efforts in consultation on transport, sustainable development and waste strategies have simply resulted in small numbers of the same kinds of people and organizations saying the same things as they would have in direct consultations, but instead over the web. Some of the big companies that have tried the same route have found that the public have not become engaged, but, unexpectedly, employees have used web consultations to raise environmental issues that they had previously found difficult to voice. The web is a useful, though very far from adequate, response to failures of participation in politics or to information deficits.

Activity 8.7

Here is set of optimistic statements about the role of Internet-based sources of environmental information. Note down your own first responses to the statements before looking to see whether you agree with my own attempt to do this activity. I suggest you spend half an hour or so studying websites to look for examples that support or undermine the statements.

(a) The web is cheap to produce and accessible to anyone, anywhere, who has an Internet connection: surely we'll all become our own environmental journalists. The Internet brings an end to editorial control over what we see as news, and we are no longer constrained by the prejudices and limitations of traditional news sources. If we think it's news, we can make it news.

(b) Governments and business can now communicate with citizens or 'stakeholders' directly about their engagement with environmental and social issues, unmediated by news media. We will be able to participate directly in environmental decision-making through new forms of electronic democracy.

(c) The web transforms the way we receive and act on information. By cross-referencing through hypertext links, by using search and archive facilities, and by sharing thoughts and questions with other citizens on newsgroups we will become more environmentally literate and motivated. A global ethic of care for environment and distant others will be underpinned by the web.

These are big issues, without right or wrong answers. My own responses to these statements are given at the end of the chapter.

Summary

In this section we have:

- looked at how new technologies and increasing competition are resulting in changes to news production that may limit and distort understanding and debate of complex environmental issues;

- noted that the convergence of media ownership into the hands of just a few vast multi-media companies engaged in intense competition may result in homogenization of the news and bias in reporting: environmental stories may find it more difficult to get space in this context, but when they do they are likely to make a bigger impact;

- found that some researchers argue that audiences are sophisticated interpreters of media texts, and that there are as many readings of a particular media text as there are readers/viewers, and hence have come to the positive conclusion that the hazards that come with narrow editorial control of news decision-making are reduced by the audiences themselves;

- recognized that Internet-related technologies and cheap digital media are bringing as many opportunities as threats in terms of enabling expression and debate;

- noted that the web also gives NGOs and the public a powerful tool for campaigning on environmental issues.

5   Conclusion

Through the course of this book you have been exploring different ways of looking at the sources and nature of environmental debates and conflicts. Power has been one of the analytical concepts that we have used to help you do this. This chapter has considered the power that media decision-makers have to dictate what we read, hear and watch, and also the power that the sources of stories have to shape the news. It has also invited you to think about how power is spread unevenly. With regard to media production, there are widely varying levels of access to journalists and editors, sometimes to do with geography, sometimes to do with political or economic power. In terms of media consumption, the fact remains that the uneven availability of technology and other social and economic factors can limit people's capacity to make the most of the different media to help resolve environmental debates and conflicts. Furthermore, the ceaseless daily

Figure 8.18 A young Haitian orphan taking the chance to harangue the media at an election rally: there are some new ways his voice might be heard.

tide of images, information and opinion about events beyond our immediate control can be disempowering.

More positively, there are ways in which the range of news media can give citizens power. They can equip people to scrutinize the work of governments and companies everywhere. Most of us rely on media channels to find out about new scientific and other understandings of interactions between the human and non-human world locally and globally. The media can make environmental change in some sense tangible. Also, audiences are not passive recipients of these messages. As the last thirty years of environmentalism has shown, people can and do act on what they learn from the news and other media, both individually and collectively, and this has fed into new ways of doing politics.

We have seen in many cases in this book that local and global environmental understanding and action are bound up together. In this chapter we have looked at the potential and limits of the media in enabling societies to think through, debate and act on what it means to live in an interconnected world. Several chapters in this book have touched on the ways in which economies, communications, environmental impacts and ethical considerations are globally interconnected. Certainly our actions have far-reaching consequences in space and time, but we are also increasingly aware of them. Just when environmental issues seem to be most daunting, we are beginning to get a sense of just how powerful we can be in responding to them.

References

Bardoel, J. (1996) 'Beyond journalism: a profession between information society and civil society', *European Journal of Communication*, vol.11, no.3, pp.283–302.

Brooks Johnson, C. (1998) *Global News Access: The Impact of New Communications Technologies*, London, Praeger.

Castree, N. (2003) 'Uneven development, globalization and environmental change' in Morris, D., Freeland, J.R., Hinchliffe, S.J. and Smith, S.G. (eds) *Changing Environments*, Chichester, John Wiley & Sons/The Open University (Book 2 in this series).

Harrabin, R. (2000) 'Reporting sustainable development: a broadcast journalist's view' in Smith, J. (ed.).

Hinchliffe, S.J. and Blowers, A.T. (2003) 'Environmental responses: radioactive wastes and uncertainty in Blowers, A.T. and Hinchliffe, S.J. (eds) *Environmental Responses*, Chichester, John Wiley & Sons/The Open University (Book 4 in this series).

McNair, B. (1998) *The Sociology of Journalism*, London, Arnold.

Natural Environment Research Council (1996) *Scientific Group on Decommissioning Offshore Structures*, April, Swindon, NERC.

Natural Environment Research Council (1998*) Scientific Group on Decommissioning Offshore Structures*, May, Swindon, NERC.

Rose, C. (1998) *The Turning of the Spar*, London, Greenpeace.

Smith J. (ed.) (2000) *The Daily Globe*: *Environmental Change, the Public and the Media*, London, Earthscan.

Snow, J. (1997) 'More bad news', *The Guardian*, 27 January.

Tuchman, G. (1978) *Making News: A Study in the Construction of Reality*, New York, The Free Press.

Urry, J. (2000) *Sociology Beyond Societies: Mobilities for the Twenty-first Century*, London, Routledge.

Answers to Activities

Activity 8.5

Table 8.3

Sources of news content	Can they put out a press release that journalists are likely to use?	Can they provide high quality images and / or video?	How likely is it that they will be asked to comment on an issue?
UN body (e.g. UNEP)	Yes. UN and governments, NGOs and business all tend to employ former journalists as press officers	To some degree, but they put more effort into providing authoritative textual information to back up press releases	Diplomatic constraints make it difficult for these bodies to say much of interest 'on the record', but they might leak interesting material that shows up, for example, differences between countries
Well-funded developed world NGO (e.g. Greenpeace, WWF or FoE)	Yes – the most effective NGO communicators work hard to do the journalists' work for them	The top performers in this arena. Some of their pictures and footage have shaped how many of us think of environmental debates and conflicts	NGO campaigners are key sources on environmental issues, and prominent in any environment journalists contact book. But they have to be careful not to 'spin' journalist contacts with deceptions; they can easily lose their privileged status

Major global company (e.g. Shell or BP Amoco)	Yes, but journalists often tend to dismiss individual companies' 'good news' press releases, and will only pay attention to those that are reactive to, for example, an NGO campaign	They try to catch up with the NGOs on this, but rarely display the same flair in their visual media	The press officers of big companies will always be seeking to balance NGO opinion with their own take on a particular issue, though they will often have to be more proactive when an issue boils up. Sometimes they'll communicate through industry associations
UK scientist attached to a government-funded lab	They might have a gift for it or been trained, but most academics don't like to see work that's taken years to produce boiled down to a few punchy paragraphs	Academics are notoriously bad at communicating their work through images. Exceptions include the ozone depletion researchers whose colourful computer images of an ozone hole in the mid 1980s helped make the story	Much depends on personal networks and whether their institution invests in press officers who can promote researchers and their work
Independent scientist based in India	See above	Almost never, though some academics and institutes in the developing world are starting to use the web to communicate their work	Some individuals find their way into western journalists' contacts books through persistence, publication in western countries and distinctiveness of argument, but the obstacles of cost and distance from the news hubs don't make this easy
Radical grassroots NGO in the developed world (e.g. Reclaim the Streets)	Decentralized organizations make this rare, and problematic for journalists ('who sent me this?' they'll ask)	They might use the web to circulate their own pictures and video, but these images are rarely picked up by the mainstream media	Journalists don't know how to represent these loose networks. They don't have leaders as such, and are difficult to classify as sources

Activity 8.6

I took on the role of an editor of a middle-market tabloid newspaper, such as the *Daily Mail* or *Express*.

1   The other papers are doing well with this, and TV news. The plucky green minnows versus the oil giant storyline and the water cannon pictures are great; let's get this into the paper.

2   Move it up to the front page. There's some real drama here, and people really care about this; with the right pictures and headline it'll add a few thousand to circulation figures.

3   'A people's victory'; make clear the paper was on the winning side – we helped make this happen etc. Splash on embarrassment to the Prime Minister – another bungle.

4   We've got to punish Greenpeace for this – this was about good science, and journalists were conned. It's not front page, but deserves good space. We should get an editorial in too about pressure groups getting too big for their boots.

5   I hear what you're saying about this being important, but this is dull dull dull. We're a pop newspaper – I'm here to give people a cracking good read not a north London seminar group topic. If this 'stakeholder dialogue' and 'transformation of business' stuff is news, I'm a banana.

News editors are trying to keep a number of things in view as they make decisions:

- Does this story meet my news values?
- Will the story help or hinder the pursuit of more readers/viewers/listeners of the sort I'm after?
- Can the story be told within the format I work in?

These represent very real constraints when it comes to telling complex or long-running stories, for example about sustainable development and business, or about climate change or biodiversity loss.

Activity 8.7

(a) Certainly there are kinds of communication within and between groups, and between the various public audiences and institutions (NGO, corporate, governmental) that are important in environmental debate, and unmediated by professional journalism. However, in general the massive and diffuse nature of information on the web seems likely only to reinforce the editorial function that a small number of professionals have for so long played in print and broadcast media. Individuals may 'lose their way on the information superhighway and feel a greater need for journalistic direction' (Bardoel, 1996, p.285). Hence journalistic skills of synthesis, balance, attracting an audience to a story, and rapid turnaround are as important in our use of web news as they ever were. We may also want to turn to trusted 'news brands' as we look out at the ocean of information on the web, with its unstable sites, and frequently un-referenced claims.

(b) There is no doubt that there is a lot of potential in using Internet-based communications to link people with institutions, and that these allow for more two-way exchange than previous media. But it is early days, and there are some hazards too. There is the danger of false participation; the claim to be engaging with stakeholders simply because people can respond to a corporate social and environmental report on a website is at best an exaggeration. Also, there is evidence that the public has little confidence in much of the environmental information provided by big institutions, making it likely that such sources will be seen as biased and manipulative. Hence participation will be small as a percentage of the population, and in general drawn from a narrow social band. Finally, where does this participation go? How can its impacts be measured? In terms of government, old-fashioned periodic voting in representative democracies is by contrast transparent, and its results have clear impacts on how we are governed. Similarly, people may still prefer to participate in business decisions about sustainable development through supporting NGO letter-writing campaigns and protests, boycotts of companies, making ethical choices in their shopping, and to get their information on these issues from news media or NGOs themselves.

(c) Perhaps some of this is already happening; certainly powerful governments and businesses that have attracted the criticism of NGOs and the public know that Internet-based media provide environmentalists with a powerful tool. They offer almost immediate global communication of text and images at low cost, and hence a new form of global surveillance by civilians that is very difficult to counter. However, precisely those technology- and education-rich societies that are acting as the most influential hubs of global environmentalism are also 'time-poor' societies; many people won't find the time to research or act on environmental or social issues. Furthermore, the Internet itself is changing, with major companies (often from non-media sectors) moving in to become big players in the provision of access and web materials; independent citizens' concerns will have to compete with an ever wider range of commercially oriented resources for our time and attention.

Conclusion

Nick Bingham

In the introduction we set ourselves the task of taking environmental contestation seriously. To try to meet this aim we used what will be a now familiar set of conceptual tools – values, power and action – to explore a series of case studies in search of answers to the following three key questions:

- How and why do contests over environmental issues arise?
- How are environmental understandings produced?
- Can environmental inequalities be reduced?

By way of a conclusion, it is worth briefly reviewing the main points to have emerged from this approach.

How and why do contests over environmental issues arise?

For Shell in 1995, as you learned in the last chapter, the North Sea appeared as an environment in which it was safe and appropriate to dispose of the Brent Spar oil installation at the end of its working life. From the perspective of others however, as you also saw, the same body of water seemed very different. For Greenpeace and their allies, the North Sea was a fragile ecosystem that could react unpredictably to a release of toxic waste and anyway not the sort of place where something should be dumped because those responsible for it could not (or would not) think of anything better to do with it. What this example illustrates particularly clearly are two things that emerged from all the chapters in the book. The first is that different individuals and groups can make sense of environments (even what might seem to be the same environment) in very different ways. The second is that such different understandings cannot in practice always peacefully coexist together. In other words, when one group feels strongly that its understanding of a given environment is undermined by that of another, contest (and perhaps conflict) is likely to be the outcome.

What is at stake in such contests varies. Sometimes at issue are *knowledges* concerning environments. These would include, for example, disagreements over who knows best about whether landscapes should be treated as having intrinsic rather than culturally specific worth (Chapter Two) or what information is required to make optimal environmental policy decisions (Chapter Seven). Other times there are contests over the *consequences* of different environmental events or processes. These would include, for example, disagreements over what is the most appropriate source of energy under contemporary conditions (Chapter Three) or the result of damming a major river (Chapter Four). Finally, there are contests over the *livelihoods* associated with different environments.

These would include, for example, disagreements over the disparities in working conditions engendered by various ways of producing and exchanging goods (Chapter Five) or how the existence of particular environmental 'bads' are often concentrated in areas occupied by particular social groups (Chapter Six).

How are environmental understandings produced?

Where do the radically different (or sometimes not that radically different) ways of making sense of environments that animate such contests come from? Here our analytical concepts, especially values and power, offer insight. Values because (as described in most detail in Chapters Two and Seven) they both guide our actions and serve as justifications for them. We are not always fully conscious of this guidance or justification, as we pick up much of our value systems without realizing it from the various social contexts that we pass through in the course of our lives. Thus values rarely appear before us as options from which we can or must choose in any simple sense. They do, however, significantly shape who we are and what we do, and so how we deal with the environments of which we are part. It is thus no coincidence that many of the contests we have encountered in the book have had at their core a disagreement between whether to value an environment or aspect of an environment in instrumental or non-instrumental terms. The debate about the trade–environment relationship examined in Chapter Five provides a classic example.

What Chapter Five also demonstrated was that not all values are equal, at least in the sense that, at particular times and places, some of the environmental understandings that they make possible turn out to be more important or influential than others. This is as much a question of power as it is a judgement of how 'good' or 'bad' these values and understandings are. Despite the challenges represented by the protests at Seattle and after, the liberal economic understanding of trade–environment relations remains the dominant approach to such issues. Effectively it is what we have described (most notably in Chapters One and Six) as a discourse, a way of thinking about the world which, if successful, becomes very difficult to step outside. Supported by the resources of power (as discussed in Chapters One, Three and Five) – which typically act to boost one particular position while at the same time working to marginalize others – certain values, understandings and, ultimately, courses of action can come to appear very 'natural' (as the example of the liberal economic trade consensus shows). Clearly therefore we must pay attention not only to how environmental understandings are produced, but how they are *re*produced.

Can environmental inequalities be reduced?

Where does that leave us with regard to our last question? On the basis of the analysis above, the prospects might seem rather bleak. What hope is there of significant change if the unequal relationships among and between both human and non-human groups are so many, and the values and powers that produce our environmental understandings are so entrenched? While this third question is not fully addressed until the next and final book in the series **(Blowers and Hinchliffe, 2003)**, there are a couple of signs to take away from this one that suggest that we need not feel totally downhearted. The first is the simple but easily overlooked point that we can only recognize environmental inequalities at all, let alone do anything about them, if environmental contestation exists. Only then do the implications of different environmental understandings come to light and provide the opportunity to remedy something that is wrong. The situation today, in which environmental contests and conflicts are better and more widely reported than ever before, at least gives us more chances to take action.

And it is action and its possibilities that provide the second reason to be cautiously optimistic. In spite of all that we have said about the influence of values and power on what we do with respect to environmental issues, influencing is not the same as determining. We discovered in the chapters that the party with the most money or the best connections does not always prevail. The pleasant surprises (and inspiration) this situation can bring should be enough in itself to protect us against an attitude of cynical resignation in the face of environmental inequalities. But if it is not, then as well as merely being critical of those actions that cause or exacerbate them, we should also take heart from those actions that have been taken in an explicit attempt to ameliorate such asymmetrical relationships.

Thinking back over what we have covered in the book, we might recall actions that have sought to reduce inequalities of access to various environmental knowledges (including the citizen juries mentioned at the end of Chapter One, the role for education and dialogue hinted at in the conclusion of Chapter Two and the democratic possibilities of new media technologies suggested in Chapter Eight). We have also dealt with actions taken in an attempt to reduce inequalities of representation when it comes to who and what counts when making decisions with environmental consequences (including the resistance of marginal communities that formed the narrative of Chapter Three, the now famous protests mentioned in Chapter Five, or the questioning of the applications of reductive cost–benefit analyses to environmental issues discussed in Chapter Seven). Finally, we have seen examples of actions taken to reduce inequalities of provision concerning the environmental conditions in which people are expected to make a decent livelihood (including the political settlements about the distribution of water resources described in Chapter Four and the legal avenues explored with partial but significant successes by the environmental justice movement as analysed in Chapter Six). This brief review of a certain class of environmental actions is not meant to imply that we face a rosy environmental future, rather that we do not have to take environmental inequality for granted. And that is another – and the best – reason for taking environmental contestation seriously.

Acknowledgements

Grateful acknowledgement is made to the following sources for permission to reproduce material within this book.

Figures

Chapter One contents page and Figures 1.14, 1.16 and 1.17: © Adrian Arbib; *Figure 1.1 collage (clockwise from top left):* © Jeremy Horner/Panos Pictures, © Alan & Sandy Carey/Oxford Scientific Films, Courtesy of Café Direct, © Jon Spaull/Panos Pictures, © Jean-Leo Dugast, © Marcelo Hernandez/Associated Press AP; *Figure 1.4:* Adapted by permission from an illustration by Joe Zeff; *Figure 1.5:* © Daily Mail; *Figure 1.6:* © The Mirror; *Figures 1.10 and 1.20:* Ordnance Survey map of Wolston, Warwickshire, with the permission of Ordnance Survey on behalf of The Controller of Her Majesty's Stationery Office, © Crown copyright, licence number ED100020607; *Figure 1.11:* Extracts from HDRA Briefing Paper: GM Trials at Wolston. Copyright © HDRA; *Figure 1.13:* Barth, F. (1991) Insects and Flowers: The Biology of a Partnership. © 1991 by Princeton University Press. Reprinted by permission of Princeton University Press; *Figure 1.18:* © Prof. Nunjundaswamy; *Figure 1.19:* © Mark Henley/Panos Pictures.

Chapter Two contents page and Figures 2.5 and 2.22: © Chris Belshaw; *Figure 2.1:* © Courtesy of the Trustees of the Victoria and Albert Museum; *Figure 2.2:* © National Library of Australia, Canberra; *Figure 2.3:* The American Geographical Society Collection of the University of Wisconsin-Milwaukee Library; *Figure 2.4:* © Andy Park/Oxford Scientific Films; *Figure 2.6:* © Museum of Art, Rhode Island School of Design; *Figure 2.7:* © National Galleries of Scotland; *Figure 2.8:* © Tate, London 2002; *Figures 2.9 and 2.10:* © The National Gallery, London; *Figure 2.11:* © Palazzo Pubblico, Siena/photo Scala/Florence; *Figure 2.12:* © Casa del Bracciale d'Oro, Pompeii/photo Scala, Florence; *Figure 2.13:* © Godfrey Goodwin; *Figure 2.14:* © Science Museum/Science & Society Picture Library; *Figure 2.15:* Courtesy of Shirley Sargent, Flying Spur Press; *Figure 2.16:* Courtesy of The National Park Foundation; *Figure 2.17:* © Lake District National Park Authority/O.S.; *Figure 2.18:* Courtesy of the Kenyan Tourist Board, London; *Figure 2.19:* © Corinne Dufka/Reuter/Popperfoto; *Figure 2.20:* © Bruce Davidson/Magnum Photos; *Figure 2.21a:* © The Sierra Club/William E. Colby Memorial Library; *Figure 2.21b:* © Anthony Dunn; *Figure 2.23:* © Science Museum/Science & Society Picture Library/NASA; *Figure 2.24:* © Christina Berlucci; *Figure 2.25:* © Cincinnati Art Museum, The Edwin and Virginia Irwin Memorial; *Figure 2.26:* © Science Museum/Science & Society Picture Library.

Chapter Three contents page: (left) © BNFL (right) © Ecotricity; *Figure 3.2a:* unknown source; *Figure 3.2b:* © Ecotricity; *Figure 3.3:* From British Wind Energy Association website – www.britishwindenergy.co.uk/map/im/map2002;

Figure 3.4: © Martin Bond/Science Photo Library; *Page 93:* Courtesy of Dr Bob Everett, EERU, The Open University; *Figure 3.5:* © David Taylor/Science Photo Library; *Figure 3.6:* G. Sinclair at Environment Information Services; *Figure 3.7:* with thanks to Paul Gipe for the data; *Figures 3.8 and 3.9:* © Mads Eskesen; *Figure 3.11:* photo © BNFL; *Figure 3.12:* Reproduced by kind permission of Hull Daily Mail; *Figure 3.13:* With thanks to Alan Hooper, Chief Scientific Advisor, United Kingdom Nirex Ltd.; *Figure 3.14a:* © John Harris/Report Digital; *Figure 3.14b:* © British Nuclear Fuels plc; *Figure 3.14c:* © British Nuclear Fuels plc; *Figure 3.15:* © Owen Humphreys/PA Photos; *Figure 3.17:* © BNFL; *Figure 3.18:* © Martin Bond/Science Photo Library; *Figures 3.19, 3.20 (top) and 3.21:* Courtesy of Andrew Blowers; *Figure 3.20 (bottom):* © Cogema/Les films de Roger Leenhardt; *Figure 3.22:* © Department of Energy/Associated Press; *Figure 3.24:* © Cogema/Les films de Roger Leenhardt; *Figure 3.25:* © Fabian Bimmer/Associated Press.

Chapter Four contents page and Figure 4.13: © Pete Oxford/Nature Picture Library; *Figures 4.2 and 4.3:* Adapted from 'The World's Water 200-2001: The Biennial Report on Freshwater Resources', by Peter H Gleick. Copyright © 2000 Island Press. Reprinted by permission of Island Press, Washington, D.C. and Covelo, California; *Figure 4.4:* © Environment Agency Picture Library; *Figure 4.7:* © John Cancalosi/Ardea; *Figure 4.8:* © Ardea; *Figure 4.9:* © Gael Cornier/Associated Press, AP; *Figure 4.15:* © Chris De Bode/Panos Pictures.

Chapter Five contents page and Figure 5.3: © Damaris Beems; *Figure 5.1:* © Ted Grudowski; *Figure 5.4:* G2 29 May 2001 © The Guardian; *Figures 5.7 and 5.8:* © World Trade Organisation; *Figure 5.9:* © P McCully; *Figure 5.10:* © The EC Audiovisual Library; *Figure 5.11:* © J. Hartley/Panos Pictures; *Figure 5.12:* © Betty Press/Panos Pictures.

Chapter Six contents page and Figure 6.11b: © Bettmann/Corbis; *Figure 6.1 collage and Figure 6.9 collage:* (top left) © Mike Dodd, (top right, middle right and bottom left) © PhotoDisc Europe Ltd; (bottom right) © Robert Glusic/Getty Images; *Figure 6.3a:* © Bettmann/Corbis; *Figure 6.3b:* © University of Buffalo Archives; *Figures 6.4 and 6.5:* © Buffalo Courier Express Library Collection, courtesy of E.H.Butler Library Archives, Buffalo State College; *Figure 6.6:* The Fuller Projection Map design is a trademark of the Buckminster Fuller Institute © 1938, 1967 and 1992. All rights reserved; *Figure 6.7:* © Erik S Lesser/ Associated Press; *Figure 6.8a:* © David Grewcock/FLPA; *Figure 6.8b:* © David Parker/Science Photo Library; *Figure 6.8c:* © Mike Dodd; *Figure 6.8d:* © North East Trees/Photo: Adan Arreola; *Figure 6.8e:* © P.W./English Nature; *Figure 6.10:* © Rex Weyler/Greeenpeace. From Michael Brown and John May 'The Greenpeace Story', Dorling Kindersley, London 1991; *Figure 6.11a:* © Jenny Labalme. From Dumping in Dixie by Robert D.Bullard, Westview Press 2000; *Figure 6.12:* © EPA/PA Photos; *Figure 6.14a:* © Adrian Arbib; *Figures 6.14b and 6.14c:* Photo by Alfredo Quarto/Mangrove Action Project © Earth Island Institute; *Figure 6.14d:* © Lambon/Greenpeace; *Figure 6.15:* © Greeenpeace.

Chapter Seven contents page and Figure 7.4: © Heidi Bradner/Panos Pictures; *p.253 (top):* © Peter Wakely/English Nature; *p.253 (middle):* © Nick Watts/English Nature; *p.253 (bottom):* © Andy Brown/English Nature; *Figure 7.1:* Adapted from Ordnance Survey 123 South Downs Way, Newhaven to Eastbourne and Ordnance Survey 124 Hastings and Bexhill, with the permission of Ordnance Survey on behalf of The Controller of Her Majesty's Stationery Office, © Crown copyright, licence number ED100020607; *Figure 7.2:* © Paul Glendell/English Nature; *Figure 7.3:* © Jacquelin Burgess; *Figure 7.5:* R. Kerry Turner (ed.) (1993) Sustainable Environmental Economics and Management: Principles and Practice, Belhaven Press; *Figure 7.6a:* © George Bernard/Science Photo Library; *Figure 7.6b:* © Mark Carwardine/Still Pictures; *Figure 7.7:* Policy Appraisal and the Environment. Crown copyright material is reproduced under Class Licence Number C01W0000065 with the permission of the controller of HMSO and the Queen's Printer for Scotland; *Figure 7.8:* © Peter Wakely/English Nature; *Figure 7.9:* © Dr Henrik Svedsater, London School of Economics; *Figure 7.10:* Reprinted from Journal of Environmental Economics and Management, Vol. 27, K. Boyle, 'An investigation of part whole biases in contingent-valuation studies, p. 72. Copyright © 1994, with permission from Elsevier; *Figure 7.11:* © Ted Lee Eubanks; *p.273 (bottom left and right)* © Winifred Wisniewski; *p.274:* © Tony Hamblin/Frank Lane Picture Agency; *p.275:* © Roger Wilmshurst/FLPA; *Figure 7.12:* The Pevensey Levels Wildlife Enhancement Scheme: Valuation Project, Information Brochure, University of Newcastle upon Tyne on behalf of English Nature; *Figure 7.13:* O'Connor, M. (2000) Natural Capital: Environmental Valuation in Europe, Policy Research Brief No.3, Cambridge Research for the Environment; *Figure 7.14:* © Neil Bennett, www.banc.org.uk.

Chapter Eight contents pages and Figure 8.8: © Dave Sims/Greenpeace; *Figure 8.1:* © BBC Picture Archive; *Figure 8.2:* © Popperfoto; *Figure 8.3:* © Sebastian d'Souza/EPA/PA Photos; *Figure 8.4:* © PA Photos; *Figure 8.5:* Photo: Joe Smith; *Figure 8.6:* © Luciano del Castillo/EPA/PA Photos; *Figure 8.9:* © Mirror Syndication International; *Figure 8.10:* © Bernd Euler/Greenpeace; *Figure 8.11 and background to Box 8.1:* © Courtesy of Maritime GMC, Stavanger; *Figure 8.12:* © Shell International plc; *Figure 8.13:* Webpage © Corporate Watch; *Figure 8.14:* © Tim Ockenden/PA Photos; *Figure 8.16:* Webpage © Independent Media Centre; *Figure 8.17:* © Harmut Schwarzbach/Still Pictures; *Figure 8.18:* © Marc French/Panos.

Tables

Table 3.1: British Wind Energy Association; *Tables 4.2 and 4.3:* Adapted from 'The World's Water 200-2001: The Biennial Report on Freshwater Resources', by Peter H Gleick. Copyright © 2000 Island Press. Reprinted by permission of Island Press, Washington, D.C. and Covelo, California; *Tables 4.4 and 4.5:* Adapted from Soffer, A. (1999) Rivers of Fire: The Conflict Over Water in the Middle East, Rowman & Littlefield Publishers, Inc.; *Table 7.1:* Policy Appraisal and the

Environment. Crown copyright material is reproduced under Class Licence Number C01W0000065 with the permission of the controller of HMSO and the Queen's Printer for Scotland; *Tables 7.3, 7.4 and 7.5:* Reprinted from Ecological Economics: The Transdisciplinary Journal of the International Society for Ecological Economics, Vol. 33, Judy Clark et al, 'I struggled with this money business...', pp. 52-54. © Copyright 2000, with permission from Elsevier.

Cover illustrations

From left to right: © Chris Belshaw; © Adrian Arbib; © Scottish Power; © Joerg Boethling/Still Pictures.

Every effort has been made to trace all copyright owners, but if any have been inadvertently overlooked, the publishers will be pleased to make the necessary arrangements at the first opportunity.

Index